The Ecology of Fire

ROBERT J. WHELAN

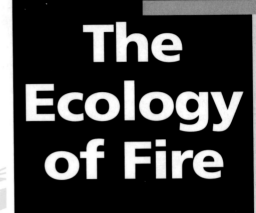

CAMBRIDGE STUDIES IN ECOLOGY

Wildfires kill many animals, but are *populations* of animals affected? How do animals survive the passage of fire? Why do some tree species survive and others die in a fire? Do frequent fires cause changes in plant community composition? How important is long-distance seed dispersal in vegetation recovery after fire? How does fire affect plant–herbivore interactions and predator–prey interactions? What are the effects of frequently applied, out of season fires for land management?

Answering questions such as these requires an understanding of the ecological effects of fire. Aimed at senior undergraduate students, researchers, foresters and other land managers, Professor Whelan's book examines the changes wrought by fires with reference to general ecological theory. The impacts of fires on individual organisms, populations and communities are examined separately, and emphasis is placed on the importance of fire regime. Each chapter includes a listing of 'outstanding questions' that identify gaps in current knowledge. The book finishes by summarizing the major aspects of ecology that are of particular relevance to management of fires – both protection against wildfires and deliberate use of fire.

The ecology of fire

Cambridge Studies in Ecology presents balanced, comprehensive, up-to-date, and critical reviews of selected topics within ecology, both botanical and zoological. The Series is aimed at advanced final-year undergraduates, graduate students, researchers, and university teachers, as well as ecologists in industry and government research.

It encompasses a wide range of approaches and spatial, temporal, and taxonomic scales in ecology, experimental, behavioural and evolutionary studies. The emphasis throughout is on ecology related to the real world of plants and animals in the field rather than on purely theoretical abstractions and mathematical models. Some books in the Series attempt to challenge existing ecological paradigms and present new concepts, empirical or theoretical models, and testable hypotheses. Others attempt to explore new approaches and present syntheses on topics of considerable importance ecologically which cut across the conventional but artificial boundaries within the science of ecology.

CAMBRIDGE STUDIES IN ECOLOGY

The ecology of fire

ROBERT J. WHELAN

Department of Biological Sciences, University of Wollongong, Australia

PUBLISHED BY THE PRESS SYNDICATE OF THE UNIVERSITY OF CAMBRIDGE
The Pitt Building, Trumpington Street, Cambridge CB2 1RP, United Kingdom

CAMBRIDGE UNIVERSITY PRESS
The Edinburgh Building, Cambridge CB2 2RU, United Kingdom
40 West 20th Street, New York, NY 10011–4211, USA
10 Stamford Road, Oakleigh, Melbourne 3166, Australia

© Cambridge University Press 1995

First published 1995

Reprinted 1997

Printed in the United Kingdom at the University Press, Cambridge

Typeset in 11/13 Monotype Bembo

A catalogue record for this book is available from the British Library

Library of Congress Cataloguing in Publication data

Whelan, Robert J.
The ecology of fire / Robert J. Whelan.
  p.   cm. – (Cambridge studies in ecology)
ISBN 0 521 32872 1 (hardback). – ISBN 0 521 33814 X (pbk.)
1. Fire ecology.   I. Series.
QH545.F5W48   1995
574.5′264–dc20   94-34787 CIP

ISBN 0 521 32872 1   hardback
ISBN 0 521 33814 X   paperback

Contents

Acknowledgements

Writing a book such as this provides an opportunity for reflection about the influences on one's research career, and an opportunity to thank those who have helped along the way. Pyrophilia has itself been a strong influence on my decisions to do research on and write about fire ecology – but the inquisitiveness to ask questions was doubtless put there long ago by my family, especially my father, whose teaching embodied the message in this quotation: *He who learns from one occupied in learning, drinks from a running stream. He who learns from one who has learned all he is to teach, drinks 'the green mantle of the stagnant pool'* [A. J. Scott, 1852].

I have been fortunate that my inquisitiveness about the workings of the natural world, especially under the influence of fire, was stimulated and honed by people who were themselves occupied in learning, particularly Don Potts, Bert Main and John Harper.

Many others also fanned my interest in linking fire and ecology, either directly or indirectly, knowingly or unknowingly – Allan Burbridge, Archie Carr, Brian Clay, Norm Christensen, Jack Ewel, Kathy Ewel, Peter Feinsinger, Malcolm Gill, Yan Linhart, Ron Myers, Peter Myerscough, Ian Noble, Bill Platt, Jack Putz, Tony Underwood, John Zasada, and Paul Zedler. Several students have increased my knowledge of fires by their research work: Jack Baker, Stan Bellgard, Ruth Ballardie, Nick de Jong, Will Edwards, Jamie Erskine, Alison Hunt, Shigeto Miyamoto, Sabine von der Burg and Pat Tap.

This book started out as a joint effort – Dr Ros Muston shared the designing of the text, the development of ideas and provided constructive criticism. Many of the approaches and ideas presented here were undoubtedly hatched in conversations with her.

John Harper's advice about embarking on a book was: 'write it *for* someone – a real person who represents the book's readership.' In a way, therefore, this book is for Alison Hunt – representing students of ecology moving from undergraduate studies to postgraduate work, and thence to research and/or management.

Many people have helped with pieces of this work as it took shape, and contributed their time, criticisms, ideas and other assistance – I am most grateful for this support – Tony Auld, Ross Bradstock, Dave Bowman, Tony Hulbert, Richard and Pat Jordan, Richard Kiltie, Paul Lefebvre, Bert and Barbara Main, Peter Myerscough, Paul Zedler. John Wiens and an anonymous reviewer read the whole manuscript and made valuable suggestions and identified countless errors. The final manuscript, figures and tables may never have been completed without the organization and assistance provided by Darien Arthur.

I acknowledge the following for financial support of my own research on fire ecology and other support in relation to the writing of this book – University of Western Australia Postgraduate Studentship, Archie Carr Postdoctoral Fellowship (University of Florida), sabbatical leave support (University of Wollongong, Fulbright Senior Award, Australian Research Council. In addition library, research and other facilities were provided by: the University of Western Australia, University of Wollongong, University of Florida, San Diego State University, Barren Grounds Nature Reserve.

Two groups of people have put up with the frustrations of being associated with me as an author. Martin Waters and especially Alan Crowden at CUP have willingly provided support, information, leniency and chiding. Last but not least, Anna and Megan provided support, an anchor, welcome distraction, encouragement and a reminder that there is more to life.

1 · *Fire ecology – an introduction*

> The biological response to a fire can vary widely. It will depend, first, on the physical properties of the fire – its intensity, size, frequency and time of occurrence – all of which influence the chemical potential for combustion and determine the nature of the chemicals liberated by combustion. It will depend, too, on the genetic potential stored within biota, which may also be released by a fire, and on the mechanisms or relationships for exploiting a fire that may exist within the biota. *(Pyne (1982) p. 38)*

Fire is a topic on which most people can comment. Fire is a widespread phenomenon. Most of us have seen fires in natural vegetation, or their effects; stark, blackened vegetation or a smoke pall. Because fires such as these can have damaging economic and social effects, can spoil forestry timber, can burn down houses and farms, and can kill people and animals, there has been a lot written about wildfires. Added to this wide perception of the damage that can be caused by wildfires, there has been increasing publicity given, since the 1950s, to the active use of fire as a management tool, particularly in protecting against severe wildfires. The introduction of a policy of deliberate burning as a management tool has a fascinating history, especially in the United States Forest Service (see Schiff 1962, Pyne 1982), but the ecological effects of prescribing a fixed burning regime on large tracts of land are increasingly being questioned.

To an ecologist, fire can be treated as just one of many factors in an environment. It compares with droughts, floods, hurricanes and other physical disturbances because of the direct impact it makes on organisms. Unlike these physical factors, however, fire as a disturbing force is itself influenced by the biota, particularly the plant community. Alteration of the vegetation by any of a number of factors can influence the nature of a subsequent fire. Fire has similarities to grazing as a force on vegetation because of such feedback effects.

Although knowledge of the ecological effects of fire has contributed to the development of ecological theory, an understanding of fire also has broader significance. First, much of the world's forestry is conducted in natural ecosystems. Ecological processes, including fire, therefore impinge on long-term forest productivity in an economic sense. Second, a substantial amount of grazing by domestic stock is carried out in grasslands, rangelands, and pastoral land. Fires in these areas affect agriculture directly by removing biomass and killing animals, and indirectly by changing plant productivity and species composition. Third, fire affects the quality and quantity of water harvested from water catchments. Fourth, an understanding of the ecological processes that permit a plant community to recover after natural disturbances, such as fire, contributes to the success of efforts made to revegetate lands following man-made disturbances. Finally, an understanding of the ecological effects of fire, both wildfire and management burning, is fundamental to conservation of plant and animal populations and representative communities in many areas.

Application of knowledge about the effects of fire to each of the management tasks listed above depends upon a sound background in ecology and access to the literature dealing with fire. There is a need for a general treatment of fire ecology for the following reasons:

1. Students embarking upon research in relation to fire should familiarize themselves with the complexities of the field, be introduced to appropriate experimental and techniques, and be made aware of the major gaps in our knowledge.
2. People working with fire as a management tool are being forced more and more to take account of the ecological implications of their management practices. For these people, this text is an introduction to what may be an unfamiliar set of ideas and literature.
3. Finally, all of us – researchers, students and land managers – need to be reminded from time to time that there is a broader perspective on ecological problems than the approach on which we usually focus.

This book is not intended to be a comprehensive review of all the detailed knowledge, from every region of the world, of the ecological effects of fires. Some of this sort of information is available in various treatments of fire ecology that have been published recently (Table 1.1). Most of these publications have a regional slant and emphasize the need for further research into various aspects of fire ecology, pointing to specific unanswered questions in each region. However, the interpretation of the *general* significance of individual studies and specific questions and the application of conclusions from one study to an

Table 1.1. *Recent books in fire ecology*

Booysen, P. de V. and Tainton, N. M. (eds.) (1984) *Ecological Effects of Fire in South African Ecosystems.* Springer-Verlag, New York.
Chandler, C., Cheney, P., Thomas, P., Trabaud, L., and Williams, D. (1983) *Fire in Forestry* (Vols. 1 & 2). Wiley, New York.
Cowling, R. (ed.) (1992) *The Ecology of Fynbos: Nutrients, Fire and Diversity.* Oxford University Press, Oxford.
Fuller, M. (1991) *Forest Fires: An Introduction to Wildland Fire Behavior, Management, Firefighting, and Prevention.* Wiley & Sons, New York.
Gill, A. M., Groves, R. H. & Noble, I. R. (eds.) (1981) *Fire and the Australian Biota.* Australian Acacemy of Science, Canberra.
Goldammer, J. G. (ed.) (1990) *Fire in the Tropical Biota.* Springer-Verlag, Berlin.
Johnson, E. A. (1992) *Fire and Vegetation Dynamics: Studies from the Boreal Forest.* Cambridge University Press, Cambridge.
Minnich, R. A. (1988) *The Biogeography of Fire in the San Bernardino Mountains of California: A Historical Study.* University of California Press, Berkeley.
Mooney, H. A., Bonnicksen, T. M., Christensen, N. L., Lotan, J. E. and Reiners, W. A. (eds.) (1981) *Fire Regimes and Ecosystem Properties.* USDA Forest Service Gen. Tech. Rep. WO-26. Washington DC.
Mooney, H. E. and Conrad, C. E. (eds.) (1977) *Environmental Consequences of Fire and Fuel Management in Mediterranean Ecosystems.* USDA Forest Service Gen. Tech. Rep. WO-3. Washington DC.
Pyne, S. J. (1984) *Introduction to Wildland Fire.* Wiley and Sons, New York.
Pyne, S. J. (1991b) *Burning Bush: A Fire History of Australia.* Holt, New York.
Trabaud, L. (ed.) (1987b) *The Role of Fire in Ecological Systems.* SPB Academic Publishing, The Hague.
van Wilgen, B. W., Richardson, D. M., Kruger, F. J. and van Hensbergen, H. J. (eds.) (1992b) *Fire in South African Mountain Fynbos.* Springer-Verlag, Berlin.
Wade, D., Ewel, J. J. and Hofsetter, R. (1980) *Fire in South Florida Ecosystems.* USDA Forest Service Gen. Tech. Rep. SE-17. Asheville, North Carolina.
Walstad, J. D., Radosevich, S. R. and Sandberg, D. V. (eds.) (1990) *Natural and Prescribed Fire in Pacific Northwest Forests.* Oregon State University Press, Corvallis.
Wein, R. W. and MacLean, D. A. (eds.) (1983) *The Role of Fire in Northern Circumpolar Ecosystems.* Wiley and Sons, New York.
Wright, H. A. and Bailey, A. W. (1982) *Fire Ecology: United States and Southern Canada.* Wiley-Interscience, New York.

understanding of fire in another region require a broad perspective. There is a great deal to be gained by contrasting different regions and one aim of this book is therefore to provide a basic framework concerning the ecological effects of fires, which can be used to direct future investigations and assist interpretation of existing information.

Although one of the aims of this book is to examine general patterns and principles in fire ecology, there is nevertheless a geographic bias produced by my own experiences. Thus, many examples are drawn from studies in south-east and south-west Australia, Florida and, to a lesser extent, California.

The material presented in this book is divided into sections describing the phenomenon of fire, the responses of individual organisms to fires, and the responses of populations and communities to fires.

Students and researchers in fire ecology too frequently treat fire as an isolated but repeatable event, without acknowledging that one fire is not like another. In contrast, most ecologists would view it as foolish to treat a single dry spell in summer as ecologically equivalent to a 5-year-long drought. Grubb (1985) discussed the importance of this feature of fire, and this view has been strongly emphasized for over a decade, notably in A. M. Gill's writings about *fire regimes* (see, for example, Gill 1975, 1981b). A fire has the immediate characteristics of intensity, season, extent and type (i.e. humus, ground, crown) and historic characteristics such as pre-fire climate, time since previous fires and characteristics of previous fires. These immediate and historic components are not independent, since the history can exert a strong influence on the immediate fire characteristics.

This situation, combined with the need to focus on fairly large areas for the study of many individual organisms, populations or communities, makes replicated, long-term fire studies difficult. 'Replication' must therefore often come from an examination of the results of a number of independently conducted studies of similar organisms in similar environments. Dissecting background variability from real effects of the fires is problematic, and as many factors as possible causing variation among studies must be identified. If a study does focus on a single fire in isolation, descriptive information about the particular fire must be collected and published. Knowing what characteristics of fire to measure and how to do it are therefore important parts of a study of fire ecology, even if an organism, rather than the fire, is the primary focus of attention. The phenomenon of fire is therefore included as a separate chapter (Chapter 2).

Armed with this background, the remaining chapters explore the ecology of individual organisms, populations and communities in relation to fire. In each section, the general nature of ecological studies at that level of organization are explored and then fire effects are examined in this ecological context.

In writing a book such as this, there are certain approaches, principles and ideas which could be emphasized. In various sections of this book, my emphases will be apparent, but they are stated explicitly here.

1. The ecological effects of fire can be extremely complex. Although experimental studies must focus on specific questions, the possibility of interaction and second-order effects must be borne in mind. In particular, plants and animals are all too often separated in ecological studies. Such an approach may miss the important effect of fire affecting, for example, an herbivore population that in turn influences the plant community.

2. Fire should not be viewed as a catastrophic event in most situations. In the context of a species, a population or a community, the response to a fire is actually a response to one of a series of recurring events. The history of past fires at a site contributes to the response to the current one. In this respect, fire may be similar to many other recurrent events affecting organisms, such as droughts, floods, cyclones and hurricanes.

3. A consideration of time-scales of fires in relation to the organisms or communities involved must underlie an understanding of the ecological effects of fire. An individual tree may see many fires in its lifetime, whereas an insect or small-mammal species may go through many generations between fires.

4. Fires in some vegetation communities may indeed be considered catastrophic because, even though fires may recur, the frequency has been so low in the past that interfire intervals have exceeded many generations of the organisms affected. Similarly, fires of unusually high intensity or completely out-of-season may have catastrophic effects even in fire-prone environments.

5. One must be wary of inferring that characteristics which permit an organism to survive a fire should be considered 'fire adaptations'. A given characteristic may have arisen in response to a selection pressure other than fire. Also, the response of a population to a single fire is not a very good estimate of the fitness of individual lineages faced with a series of fires.

6. Perhaps too much emphasis is placed on obtaining an assessment of the so-called 'natural' fire regime of an area. It is frequently stated that, in the absence of more precise information about the ecological effects of particular fire regimes, the best management plan is to mimic nature. Although attempts can been made to estimate historic fire frequencies over various time spans (e.g. pre-human, pre-

industrial) and to include estimates of the areas burned in past fires, other components of so-called 'natural' fire regimes, such as fire intensity, season and type (i.e. crown versus ground) are much more difficult to estimate, though equally important. Furthermore, even without the assistance of modern *Homo sapiens*, fire regime is sure to have varied widely over time, and the variation is likely to have been ecologically significant in itself. In the long run, a profitable approach may be to place more emphasis on understanding just how organisms, populations and communities respond to experimentally imposed fire regimes. Anthropogenic ignitions, increasing dissection of natural landscapes and reduction of fire-hazard are realities, and direct investigation will therefore become more and more valuable than inferences based solely on the estimation of natural fire regimes in the pre-human past.

7. There are very few long-term, experimental studies of the effects of fires on any level of organization – individual organism, population or community. Long-term studies are an absolute necessity for a variety of reasons. First, many organisms in fire-prone ecosystems have long lifespans, and the effects of even a single fire may not be revealed for many years. Second, management practices are based on assumptions that certain long-term processes are operating in the community. The validity of these processes has rarely been tested by long-term, empirical study. The need for validation of models of long-term change based on short-term studies is becoming urgent. Third, ecological effects of fire are responses to a sequence of fires occurring with a certain range of frequencies, intensities, seasonalities and extents. Let us focus on any one of these, say frequency, and hold the others constant in an experimental design. An experimental comparison of low-frequency fires (i.e. one fire in 50 years) and high-frequency fires (i.e. one fire in 5 years) will clearly take many decades. For researchers in short-term positions, these comments should not be disheartening. It will be clear from the material discussed in this book that in fire ecology there is also much need for, and scope for, carefully designed and conducted shorter-term experiments.

8. My approach is largely mechanistic, in the sense that much of ecology is seeking explanations for why observed patterns are the way they are. Hence, my emphasis is on studies that are experimental and comparative, rather than solely descriptive. There is a large literature describing patterns, especially of vegetation, in relation to fire (see, for example, many of the reference books listed in Table 1.1) that is not covered thoroughly in this book.

Some definitions

In a field about which so many people know at least a little, it is as well to consider fire terminology that differs from one place to another. Terms used to describe fires include:

wildland fire	forest fire	prescription fire
bushfire	scrub fire	controlled fire
wildfire	brush fire	hazard-reduction fire
firestorm		cool-season fire

All this can be very confusing, and the terms variously describe (i) the intensity of a fire, (ii) whether it was planned by some land management authority, and (iii) the type of vegetation in which it occurs, or some combination of these three.

The *bushfires* of Australia include fires in forests, heaths and grasslands, and this term is probably equivalent to the term *wildland fire* that is commonly used in North America. These are generic terms that describe fires in natural vegetation, but typically they are fires that were not planned, but were started by arson, accident or lightning and burn out-of-control. The term *wildfire* certainly describes a fire that is out of control.

Terms such as *forest fire* and *brush fire* typically describe the vegetation type in which a fire occurs.

Fires are deliberately lit by land management agencies for a variety of reasons. These will usually be constrained by some *prescription* that is written down and considered prior to burning. Reasons for prescribing fire include *hazard-reduction* (that is, removing some biomass so as to reduce the intensity of a future wildfire) and these fires are typically 'controlled' and, to achieve this, conducted in a *cool season* of the year. High-intensity fires may also be prescribed, at times, to maintain a vegetation type, remove invasive weeds or manipulate faunal habitat. These would also be *controlled*, though not necessarily in the cool season.

There is another suite of terms that describe some characteristics of a fire. These include: head-fire, back-fire, back burn, burning off, burning out. The first two describe the position of a fire front relative to the wind: a head-fire burning with the wind and a back-fire burning into it. Back burning, burning off and burning out are three terms that describe deliberate burning, usually conducted during fire-fighting operations.

2 · Fire – the phenomenon

This chapter examines the physical and chemical characteristics of fire. A comprehensive understanding of fire at this level of detail is important for ecologists because of the two-way interactions between the characteristics of the vegetation and the nature of a fire (Fig. 2.1). The physical and chemical reactions of combustion determine the underlying nature of a bushfire; whether it ignites, how hot it burns, how it behaves. The likelihood of a fire starting, given that an ignition source is present, may depend largely on physical and chemical factors. Variability in aspects of fire behaviour such as intensity and rate of spread may be determined, to a greater or lesser extent, by factors influencing the basic physical and chemical reactions.

An understanding of these basic principles will also reveal how the existing vegetation can influence the nature of fire at a site. The characteristics of the fire, with all these interacting causes, will influence strongly the responses of particular plant and animal species (to be considered in later chapters). The responses to fires of any particular element of the biota are highly variable, making predictions difficult. It is therefore very important that the variability among individual fires is at least described adequately in experimental, ecological studies, even if fire characteristics cannot be precisely controlled.

There are several excellent book-length treatments of the physical characteristics of fire and of fire behaviour. In particular, Luke and McArthur (1978), Chandler et al. (1983; Vol. 1), Pyne (1984) and Johnson (1992) are valuable references that elaborate some of the information presented in this chapter. It is useful to recognize that our understanding of the factors determining fire behaviour has grown through a series of models of increasing complexity. Thus, a simple fire-triangle model, which links fuel, moisture and oxygen as primary determinants of fire intensity, has developed through relatively simple empirical models (McArthur 1966, Noble et al. 1980) into more complicated, computer-aided predictions, such as BEHAVE (Burgan

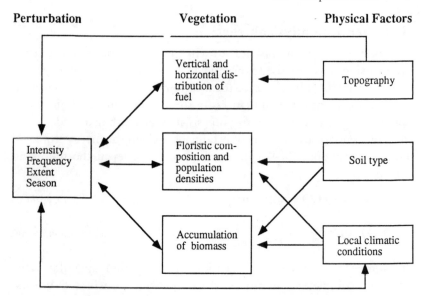

Figure 2.1 Schematic diagram illustrating the interactions between characteristics of fire, the vegetation and physical conditions (modified from Riba and Terridas 1987).

and Rothermel 1984, Catchpole and de Mestre 1986). In general, these models are based both on factors regulating the basic combustion reactions and also on an understanding of the ways in which various characteristics of climate, plant communities and the physical environment influence ignition and fire behaviour.

It is not within the scope of this book to explore these models: detailed treatments may be found in Burgan and Rothermel (1984), Catchpole and de Mestre (1986), Chandler *et al.* (1983) and other works. However, it is important for ecologists to understand combustion, and to be convinced that a description of relevant physical measures of fire is necessary in any ecological study.

In the field, many features of fuel, weather, topography and fire history interact in various ways to modify the *potential* fire characteristics, determined ultimately by the amount of energy stored as live and dead biomass per unit area. The following sections examine these interactions. The occurrence of fire is separated into two components, namely the initial ignition, or the establishment of flaming combustion, and the subsequent spread. Different, though overlapping, groups of factors control each.

The chemical–physical reaction

The basis of fire is the physics and chemistry of combustion. Energy stored in biomass is released as heat when materials such as leaves, grass or wood combine with oxygen to form carbon dioxide, water vapour and small amounts of other substances. In some ways, this reaction can be thought of as a reverse of photosynthesis, in which carbon dioxide, water and solar energy are combined, producing a chemical energy store and oxygen. Trollope (1984) compared these processes in the following equations:

$$CO_2 + H_2O + \text{Solar energy} \rightarrow (C_6H_{10}O_5)_n + O_2 \quad \textit{Photosynthesis}$$

$$(C_6H_{10}O_5)_n + O_2 + \frac{\text{kindling}}{\text{temperature}} \rightarrow CO_2 + H_2O + \text{heat} \quad \textit{Combustion}$$

In a simple form, the chemical equation for combustion can be illustrated with the complete combustion of a simple sugar, such as D-glucose (McArthur and Cheney 1972). In this case:

$$C_6H_{12}O_6 + 6O_2 \rightarrow 6CO_2 + 6H_2O + 1.28 \times 10^6 \text{ kJ}$$

Plant material is, of course, chemically much more complex than glucose, and different components of fuel (e.g. dead leaf litter, dead wood, live foliage, twigs and wood) have energy stored in a great variety of forms. However, the principle of the equation is much the same.

The term 'kindling temperature' in the above combustion equation indicates that combustion is neither a simple nor a spontaneous process. It requires the 'activation energy' of an external energy source. The process of combustion may be broken down into several stages. Wood is set on fire by first bringing to bear enough heat, from an external source, to cause pyrolysis. The colder the fuel is initially, the more kindling energy is required. Pyrolysis is the thermal alteration of the fuel, resulting in the release of water vapour, carbon dioxide and combustible gases, including methane, methanol and hydrogen. During pyrolysis, the reaction changes from being exothermic (i.e. requiring heat to proceed) to being endothermic (self-sustaining). Applying a pilot flame to the combustible gaseous products escaping from the wood and mixing with air during active pyrolysis results in flaming combustion.

Three stages of combustion in a vegetation fire can be recognized in relation to the basic principles of combustion. These are: (i) preheating, in which the fuel just ahead of the fire front is heated, dried and partly pyrolysed; (ii) flaming combustion, which results from the ignition of

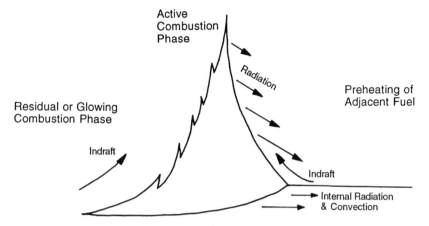

Figure 2.2 Flame profile of a fire on flat ground with no wind, indicating the region of pre-heating, flaming combustion and glowing combustion (modified from Cheney 1981, Rothermel and Deeming 1980 and Alexander 1982).

the flammable hydrocarbon gases; and (iii) glowing combustion, during which the remaining charcoal burns as a solid, with oxidation taking place on the surface leaving a small amount of residual ash.

An understanding of the processes of pyrolysis and ignition shows how a fire can be thought of as a chain reaction, with the initial ignition source providing the activation energy that permits ignition and self-sustainability of a fire. Flaming combustion at the fire front then preheats adjacent fuel and provides the pilot flame to cause its ignition (Fig. 2.2).

The amount of energy produced by the combustion reaction is of particular interest because it is strongly linked with a number of measures of fire intensity. The heat yield of a combustion reaction is related to the total energy that would be released by complete combustion of the fuel, modified by various factors. First, the total heat of combustion values vary only slightly among different fuels but may be influenced strongly by the presence of volatile oils and resins that have higher energy contents (Table 2.1). For example, Pompe and Vines (1966) found that eucalypt leaves oven-dried at 110 °C yielded less heat energy than would have been expected from a previous study (Walker 1963) of the heat contents of their distillation products and residues (16.74 versus 20.93 MJ/kg). Pompe and Vines attributed this difference to loss of the volatile oils distilled off during drying (see also King and Vines 1969).

Table 2.1. *Total heat of combustion values for a variety of Australian and North American forest fuels. These are values that would be obtained by complete combustion in a bomb calorimeter*

Fuel type	Heat of combustion (MJ/kg)
Woods	
Oak	19.33
Beech	19.98
Pine	21.28
Poplar	18.22
Eucalyptus capitellata	19.92
E. viminalis	19.54
E. melliodora	19.64
E. obliqua	19.23
E. macrorrhyncha	19.27
E. rostrata	20.45
E. polyanthemos	20.20
E. elaeophora	19.91
E. amygdalia	21.35
Acacia melanoxylon	18.88
Other fuels	
Pine sawdust	21.74
Spruce sawdust	19.65
Wood shavings	19.18
Pecan shells	20.68
Hemlock bark	20.35
Pine pitch	35.13
Eucalyptus oil	37.20

Source: McArthur and Cheney (1972).

Second, even with complete combustion of the fuel, heat is lost though the vapourization of water in the fuel. Heat is used in: (i) raising fuel water temperature to 100 °C; (ii) separating bound water from the fuel; (iii) vapourizing the water in the fuel; and (iv) heating the water vapour to flame temperature. This 'loss' of heat is important because it determines the transition from an exothermic to an endothermic reaction during pyrolysis, and thus will partly determine whether an ignition source can start a fire. Starting a campfire with wet twigs is difficult because a great deal of the heat given off by the burning match is taken up heating the water in the twigs. Data obtained by McArthur and

Cheney (1972) suggest that the energy lost due to vapourization of water is a linear function of the moisture content of the fuel. Although it may be important in determining initial ignition, Vines (1981) pointed out that this 'lost' energy is a very small proportion of the heat content of the dry fuel (only about 10% at a moisture content of 50%), so it should be insignificant once the endothermic reaction is under way. Evidence of this is seen in the vast amounts of water that must be sprayed into a burning house to bring the fire under control. A much smaller amount, applied prior to ignition, would have prevented the fire taking hold in the first place.

Finally, incomplete combustion of the fuel will, of course, reduce the energy output below the potential maximum. Incomplete combustion describes two possible processes. First, all components of the burning fuel may not be converted to released energy, either being given off as particulate carbon compounds in the pall of smoke accompanying the fire or remaining as charred plant material at the site. Second, not all the biomass at the site may ignite in the first place.

Temperature profiles

The intensity of a fire varies both horizontally and vertically, determined partially by the distribution of the fuel. Thus, an estimation of *fireline intensity* (energy output per metre of fire front), which is determined in part by the amount of fuel available at a given location, may obscure ecologically important variation in peak temperature and duration with height and over space. It is important to note that variation in two components of intensity, namely peak temperature reached and duration of a given temperature, may differ markedly among fires. Thus, a peat fire may smoulder for many hours or days at a particular spot, not reaching particularly high surface or sub-surface temperatures but sustaining a lethal temperature for the whole time. The vertical distribution of temperature, both above and below ground, is of particular significance to the survival of individual plants and plant parts.

Above ground

The vertical distribution of temperature in a fire is determined by several factors, including distribution of the fuel, wind speed and direction of the fire front (i.e. head-fire or back-fire). Most measurements have recorded only peak temperature and not duration of temperature, although both are important for survival of plant tissues (see Chapter 3).

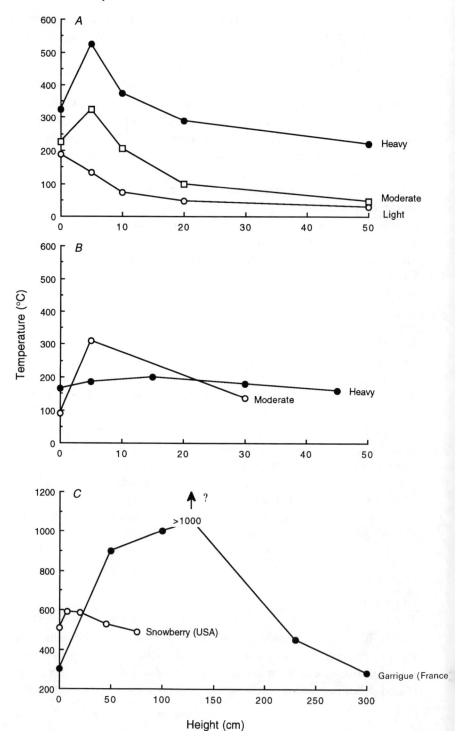

Above-ground temperature eventually declines with height (Fig. 2.3), as the heat generated by combustion is dissipated in an ever-increasing volume of air. Stronger winds during a fire therefore may result in lower temperatures at a given height than if the fire were burning in still air. Looking more closely, the decline in temperature with height is not simple and may vary substantially among fires. Some studies suggest that the maximum temperature increase occurs at some distance above the ground (Fig. 2.3), perhaps at the top of the vegetation layer constituting the fuel in a particular fire. The reasons for such variations among fires are not yet clear, but the temperature profile is most likely to be a combination of convective transfer of heat upwards and the vertical distribution of fuels.

Spatial heterogeneity of the vertical temperature profile has not been thoroughly explored, although it is important in understanding differential survival of individual plants or plant species in a particular fire (e.g. Williamson and Black 1981). Data describing mean maximum temperatures (e.g. Fig. 2.3) obscure this heterogeneity. One instructive approach was used by Trollope (1984), who obtained a frequency distribution of peak temperatures from thermocouples placed at several locations at three heights in grassland (Fig. 2.4). This study indicated that the highest temperatures in both head- and back-fires occurred at the level of the grass canopy. At ground level, there was a greater range of temperatures in head-fires, with more locations experiencing low peak temperatures but a few experiencing very high peak temperatures. The level of heterogeneity apparent in the ground-level samples was not paralleled at 1 m, suggesting that the variation in temperature at ground level 'averages out' by the time the heat–pulse reaches 1 m.

Figure 2.3   The precise relationship between peak fire temperature and height above the ground varies among studies, according to fuel load, vegetation structure and other factors, as illustrated in this small sample of studies. Others are summarized in Daubenmire (1968) and Wright and Bailey (1982).
A. Three fuel loads in grassland/shrubland (from Smith and Sparling 1966): heavy (2360 kg ha^{-1}) – solid circles; moderate (1780 kg ha^{-1}) – open squares; light (560 kg ha^{-1}) – open circles. Note that only at the lowest fuel loading did the peak temperature occur at ground level.
B. Two grassland sites: moderate fuel load (3933 kg ha^{-1}) – open symbols (Ito and Iizumi 1960); heavy fuel load (5080 kg ha^{-1}) – solid symbols (Bailey and Anderson 1980).
C. Two shrubland sites: snowberry stand in a cool spring fire (Bailey and Anderson 1980) – open symbols; French garrigue community (Trabaud 1979) – solid symbols. In the garrigue, 130 cm (height of peak temperature) represented the top of the canopy.

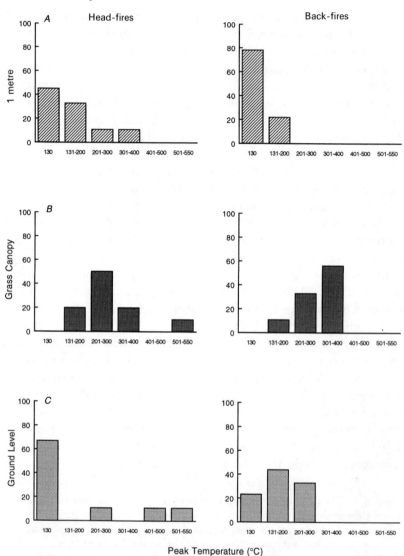

Figure 2.4 Spatial heterogeneity of the height – temperature profile in a fire is obscured in data such as those illustrated in Fig. 2.3. These histograms are frequency distributions of peak temperatures recorded by thermocouples placed at three heights in grassland: 1 m (*A*), top of grass canopy (*B*) and ground level (*C*) (data from Trollope 1984). Head-fires (left) were compared with back–fires (right). Highest temperatures were generally found at the level of the canopy.

A further complication arises in the transition from a ground- to a crown-fire (Ashton 1986). The profile of temperature and wind in the canopy during a crown fire is not known in detail, but it is likely to be very complex. The temperatures experienced in the canopy, if leaves and volatile oils are burning, must be very different from the situation in which only the ground vegetation is producing the heat which bathes the canopy. Ashton (1986) reported on the analysis of news film of the serious 'Ash Wednesday' fires in south-eastern Australia in 1983, which shows that the residence time of flames in tree crowns varied from 10 to 12 seconds and that in bushes from 5 to 35 seconds. Moreover, observations of the shape of flames indicate that strong downdraughts of cooling air come in behind the fire front to feed the strong updraughts at the front. Thus, a canopy may experience intense heat but for a relatively short time-span.

Below ground
Surprisingly little is known about the relationship between fire temperature and depth in the soil profile. It is often stated that soil is a particularly good insulator, and reports of early studies in which peak fire temperatures were measured at different depths expressed surprise at the low penetration of heat downwards (e.g. Beadle 1940). Various studies have indicated that peak temperature at 2.5 cm depth is likely to be well below 100 °C (Fig. 2.5), even when the fire above is of very high intensity. Of course, one reason for poor penetration of heat is that convective heat transfer is upwards, and radiant transfer is very transient at a given point.

Soil temperatures can be raised substantially when very high levels of ground fuel permit long residence times of fire over a particular site, such as occurs during the burning of post-logging slash (Fig. 2.5 C, D). An interesting complexity to the relationship between fuel load and soil temperature profiles is illustrated by a study of grassland soils, by Bentley and Fenner (1958). A high fuel load (> 1 cm in this study) of litter does not necessarily ensure greater heating of the underlying mineral soil. Only the surface of the litter burned, under conditions producing a fast-moving but low intensity fire; the remainder of the litter layer insulated the soil, producing *lower* peak temperatures at the surface of the soil than occurred in sites with litter layers < 1 cm deep.

Soil moisture appears to have some influence over the dynamics of heat transfer (Heyward 1938). Moist soil reached a higher peak temperature than air-dry soil at a given depth and reached that peak more rapidly. In apparent contradiction to these data, Beadle (1940) found that

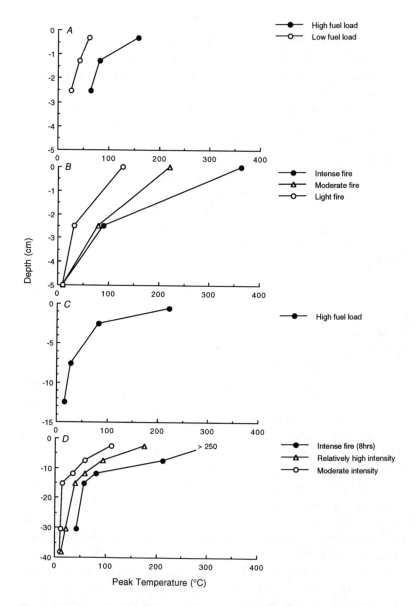

Figure 2.5 Sample of studies investigating the relationship between peak temperature and depth in the soil. In all cases, peak temperature declines very rapidly with depth.

A. Fires in longleaf pine (*Pinus palustris*) forest in south-eastern USA (from Heyward 1938).

B. Fires in Californian chaparral (from DeBano *et al.* 1977).

C. Temperatures under heavy slash fuels after logging in forest (from Neal *et al.* 1965).

D. Fires in eastern Australia eucalypt forest (from Beadle 1940).

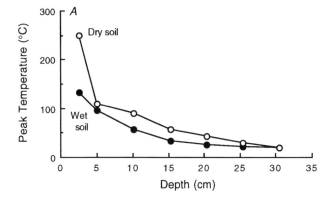

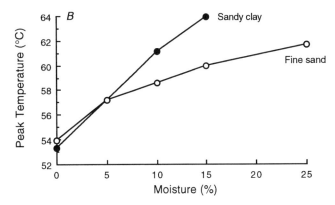

Figure 2.6   *A.* Different soil conditions can produce different relationships between temperature and depth. At all depths, dry soil (open symbols) experienced higher peak temperatures than wet soil (solid symbols) (data from Beadle 1940).
B. The effects of moisture may differ according to soil type. At soil moisture levels > 5%, peak temperature in fine sand (open symbols) was lower than that in sandy clay (solid symbols) (data from Heyward 1938).

moist soil retarded heating (Fig. 2.6). The resolution of these opposing results may lie in the temperatures applied at the soil surface. When the surface temperature does not reach 100 °C, water in the soil may facilitate conduction of heat. When the surface temperature exceeds the boiling point of water, evaporation in moist soil will delay heating of the underlying soil (Trollope 1984). Perhaps the greatest ecological significance in soil moisture lies in the differential responses of plant parts, such as seeds, and animals in burrows to moist and dry heat (see Chapter 3).

Fire temperatures and nutrients

One consequence of high temperatures generated in fires is loss of nutrients that were formerly bound in various parts of the ecosystem – soil minerals, organic matter in soils and litter, and live biomass. Losses occur by volatilization (e.g. nitrogen) and other processes such as export of fine ash in the smoke column (e.g. phosphorus). The extent of losses of nutrients appears to be related to fire intensity, with greater losses occurring in hotter fires (see Vol. 1, Ch. 7 of Chandler *et al.* 1983). The nutrient dynamics in burned ecosystems are apparently very complex and detailed exploration is beyond the scope of this book. However, the summary by Chandler *et al.* (1983) indicates that, although there will generally be some loss of nutrients that is greater in hotter fires, these losses are not always reflected in lower levels of nutrients available in post-fire soils. Fire appears to make some nutients more available by altering soil pH, and to make others more available by mineralizing them and redistributing them from biomass and necromass to the soil.

Physical characteristics of a burnt environment

The physical conditions prevailing after fire are usually very different from the pre-fire situation. Many of these differences will have profound implications for the recovery of the biota, yet relatively little information exists in the published literature. Studies by Old (1969), Knapp (1984), Mallik (1986) and Ewing and Engle (1988), in heathland and grassland communities, provide much of the available information. Similar studies are needed in a range of other plant communties. In summary, these studies indicate that fire increases maximum temperatures at the soil surface, light intensity, wind speed (below about 1 m in grassland), and the vapour-pressure deficit.

Soil temperatures

Daily soil temperatures are usually altered after fire (Fig. 2.7), with potential consequences for plant productivity, for activity of soil organisms (microorganisms are of particular importance), and even for the depth of the permafrost in Arctic areas. These effects are caused by a combination of several factors, including the removal of shading usually provided by the vegetation, the removal of the insulative effect of litter (Ahlgren and Ahlgren 1960, Old 1969), and the altered albedo of the soil surface (van Cleve and Viereck 1981).

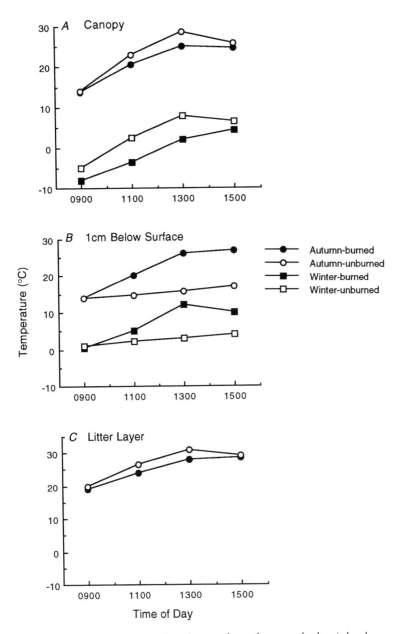

Figure 2.7 Air temperature in a site can depend upon whether it has been burned (from Ewing and Engel 1988), burned sites being generally cooler at canopy level (*A*) but warmer in the soil 1 cm below the surface (*B*), with little difference in the leaf litter/grass layer (*C*). Air temperatures were measured in burned sites (solid symbols) and unburned sites (open symbols) in autumn (circles) and winter (squares) in North American prairie.

One potentially important effect of fire on soil that has been little studied is the sterilization of the soil surface layer by high temperatures. This process has been drawn into an hypothesis about the flush of germination which follows fire in many plant communities. It is argued that allelopathic chemicals in some soils, particularly California chaparral, inhibit germination of some plant species (Muller *et al.* 1968, Wilson and Rice 1968) and that high soil temperatures during fires destroy these compounds (Christensen and Muller 1975). It has been suggested that this effect also occurs in Australian eucalypt forests (Ashton 1970) and in bracken fern stands in South Africa (Granger 1984).

Wind speed

The removal of all or some of the above-ground biomass by fire affects the wind profile in an area. Old (1969) measured wind speeds at different heights of prairie vegetation last burned at different times prior to the study. At 200 cm, the top of the grass sward, average wind speed was equivalent in recently burned and 3-year-old sites. However, at 100 cm and below, the grass of the unburned prairie reduced average wind speed to less than half that in the burned site.

It is perhaps surprising that studies like this one do not appear to have been conducted in other, more complex vegetation types. The ecological consequences of increased wind speed after fire include more rapid desiccation of potential animal habitat, plant parts and soil, greater potential for erosion and greater dispersal of some seeds.

Water relations

The removal of above-ground and surface biomass, and the increased temperatures and wind speeds in burned sites would be expected to alter the water relations of the site (see Chapter 3 of Wright and Bailey 1982 for a summary of these effects). After fire, mineral soil may be exposed to the direct impact of raindrops (e.g. Boyer and Dell 1980) and the absence of litter and surface vegetation increases the amount of surface runoff. Moist soils would then experience increased evaporation due to higher temperatures and greater wind speeds. The moisture-holding capacity of the soil is also enhanced by the amount of organic matter. Fire reduces the content of organic matter and therefore decreases the moisture-holding capacity (Neal *et al.* 1965). Mallik (1986) found changes in moisture conditions near the soil surface due to fire in Scottish heathland. Water content of the top 2 cm of soil generally decreased in response to burning, especially in summer months. Evapotranspiration was also lower in burned sites.

One effect of fire in a limited range of environments is apparently the production of a water-repellent layer in the soil (Wells *et al.* 1979). This occurs in some shrub communities (e.g. California chaparral; DeBano *et al.* 1976) when the surface fire lasts for 5–25 min. and temperatures exceed 100 °C – sufficient to cause distillation of volatile organic substances. Under these conditions, organic substances can distil downward, forming a non-wettable, hydrophobic layer in the soil (Wright and Bailey 1982).

Runoff and erosion
Reduced water holding capacity and exposure to the impact of raindrops combine with the removal by fire of litter and live plant material to expose burned sites to soil erosion. Swanson (1981) summarized the possible effects of fire on soil and geomorphic processes (see Fig. 2.8). There are some awe-inspiring examples of erosion in areas of steep, erodible soils, especially in the western United States (see photograph in Vol. 1, Ch. 7 of Chandler *et al.* 1983). In other sorts of soils, such as sandstones, fires can be an agent of weathering – causing expoliation, pitting and fracturing (Adamson *et al.* 1983, Selkirk and Adamson 1981).

At a smaller scale, surface flow of water can redistribute mineral nutrients and organic matter according to topography. Micro-terraces and small depressions act to concentrate charred organic matter and entrap seeds released after fire (see Whelan 1986, Adamson *et al.* 1983, Enright and Lamont 1989, Lamont *et al.* 1993).

Ignition in the field

Fires are not equally common nor equally devastating in all parts of the world. Temperate forests across Europe may experience a relatively large number of fires each year (167 fires per 10^6 ha) but these are usually small (average of 0.97 ha) and the estimated return period for fire at a given site is 6000 yr (Chandler *et al.* 1983). Contrast these data with the Alaskan taiga, in which there are relatively few fires per year (2.7 per 10^6 ha) but they average 1800 ha in area! The wet eucalypt forests of southeastern Australia have an intermediate occurrence of fires (66 per 10^6 ha) and an intermediate fire size (165 ha) and a return period of 43 years. This small array of statistics makes an important point – our own experience of fires is not likely to be in any way representative of what happens in other ecosystems. The extreme variation illustrated here also requires explanation. What factors might explain different levels of fire occurrence and different magnitudes of burning in different regions?

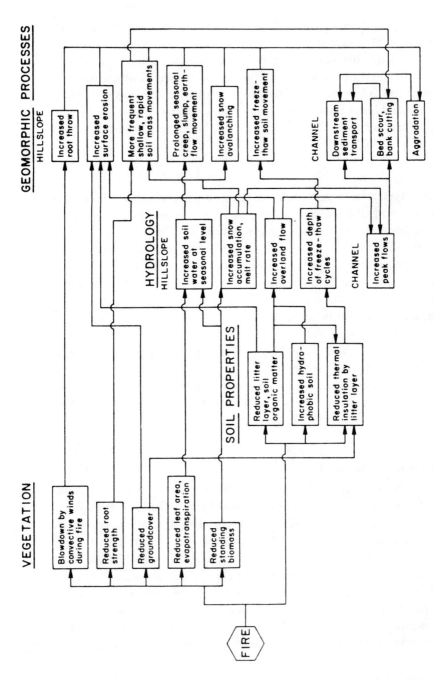

Figure 2.8 Effects of fire on soil properties, hydrology and geomorphology, both directly and indirectly – through effects on vegetation (from Swanson 1981).

Sources of ignition

An important source of ignition of wildfires in most regions of the world is undoubtedly human activity. Statistics in most areas point to anthropogenic fires, whether deliberately lit (i.e. arson, prescription fires), or accidental (including burning-off in inappropriate weather conditions, fires following vandalism, glowing cigarette butts thrown from car windows, etc.), being the most common (Komarek 1965, 1968, Gill 1981d, Kruger and Bigalke 1984).

Fires associated with human activities have a long history in many places. There are numerous reports by early European explorers of native-set fires in Australia, Africa, North and South America, and New Guinea (Mitchell 1848, Stewart 1956, McArthur 1970), and much anthropological evidence suggests that fire has long been actively used in these regions for hunting and clearing undergrowth (see references in Stewart 1956, Merrrilees et al. 1973, Hallam 1975, Nicholson 1981, Hall 1984). The anthropogenic use of fire has certainly had a long history in the Mediterranean region as well (Le Houerou 1974), reaching a peak in recent centuries due to high population densities relative to those in the New World.

Apart from anthropogenic ignitions, other sources of fire include volcanic activity, sparks from rocks in landslides, and lightning. Although the former sources of ignition may be important in localized situations (Cope and Chaloner 1985), lightning is by far the most general and widespread, non-human cause of ignition of wildfires (Stewart 1956, Komarek 1965). This point has been strongly emphasized by E.V. Komarek in a series of papers published in the *Proceedings of the Tall Timbers Fire Ecology Conferences* in the 1960s and 1970s. Although lightning has been the main non-human means of ignition for a long time in many regions, one must bear in mind that a given tract of land is currently exposed to a wide range of potential ignition sources. Ignitions associated with European settlement in North America and Australia have been common for 200 to 300 years, and aboriginal use of fire may extend back on the order of tens of thousands of years.

The underlying, non-human potential for ignition of bushfires in a region may be calculated if estimates of frequencies of occurrence of the various different ignition sources could be made. One estimate of lightning frequencies that can be easily made from meteorological records is the mean number of thunderstorm days per year. Maps showing the distribution of different thunderstorm frequencies illustrate that this ignition source is not uniformly distributed over the globe, nor indeed over any continent or smaller region. Local climatic records from any meteorological station make it possible to estimate the frequency of days, over some past period, on which lightning ignitions would have been possible, all other factors being suitable (see Figs. 5.5 and 5.6 of Pyne 1984). Regions of highest thunderstorm frequency do not necessarily all suffer high frequencies of lightning-started fires.

It is much more difficult to collect data on the frequency of additional ignition sources, independently .of whether they eventually become fires. Arson fires and accidental fires are usually only recorded by appropriate authorities as 'fires' not as 'attempts to light fires'!

Likelihood of ignition

The occurrence of an ignition source is clearly not the only determinant of the occurrence of fire. Although ignition is a necessary precursor to fire, fire is not a necessary consequence of an ignition source. A good illustration of this is the phenomenal number of lightning strikes during electrical storms in the south-east and south-west United States (see Fig. 4 of Komarek 1964). Most fail to start fires. Indeed, there is only a loose relationship between the frequency of thunderstorm-days per month and the frequency of lightning-started fires (Fig. 2.9). The explanation for this lies in the understanding of the physical and chemical reactions described above. Air temperature and moisture, fuel temperature and moisture, oxygen and suitable fuel all need to be at appropriate levels to permit a fire to start. Thus, an ignition source such as lightning that occurs when relative humidity is high, ambient temperature is low and leaf litter is saturated will rarely start a fire. This is self evident. Fear of bushfires rarely accompanies electrical storms in winter in northern England, nor indeed intense frontal storms in winter in Mediterranean-type climates.

In Mediterranean-type climates, lightning strikes that are likely to cause ignition are usually associated with convectional storms, which occur when unstable air masses are heated over hot land. These storms occur most commonly toward the end of summer and into autumn, when the long period of hot, dry conditions over summer has dried the vegetation, making ignition more likely. Thus, the seasonality of appropriate climatic conditions is the primary determinant of the season of ignition of lightning-caused fires. Similarly, unstable summer convection systems generate lightning storms in subtropical and tropical climatic regions.

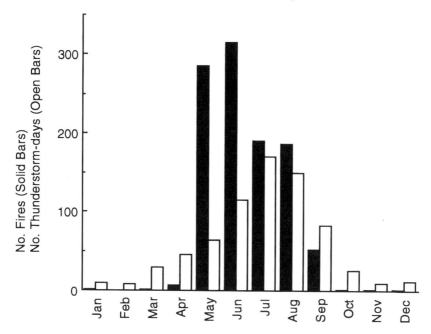

Figure 2.9 Within a single region, the timing of highest frequencies of lightning fires does not match precisely the timing of frequent thunderstorms (from Komarek 1964).

The state of the fuel at the point of application of an ignition source is an important regulator of ignition and a variety of factors controls fuel temperature, fuel moisture, and oxygen availability in the field. Ignition in Australian eucalypt forests is said to be unlikely at a fuel moisture level exceeding 20% of oven-dry weight (Foster 1976, Cheney 1981), although other fuels with high resin contents (e.g. pine) may ignite at moisture contents up to 35%. Other chemical factors may also influence the flammability of fuel. Studies of leaves of Australian trees indicate that a high mineral content reduces flammability (King and Vines 1969, Cheney 1981). Although there appears to be little variability in flammability due to mineral content among eucalypt species, dry leaves of many mesophytic species apparently possess high levels of minerals and are therefore less flammable (King and Vines 1969).

The climate in the preceding days and at the time of ignition will be the primary determinant of the fuel's moisture content. The nature of the vegetation at the site can interact with climate because of the influence of the vegetation over the microclimatic conditions at the surface. A dense, closed forest canopy will reduce wind speeds and sustain high humidity at the surface for some time after ambient conditions become hot and dry. These points have been well illustrated recently in a study by Uhl and Kaufmann (1990) of the potential for wildfire in disturbed Amazonian rainforests. For 2 weeks after a substantial rainfall, measurements were made of changes in moisture content of experimental 'fuel sticks' (1 cm diameter sticks suspended 25 cm above ground) and bagged litter samples on the ground. Logged forest and pasture dried out more rapidly than primary forest. During the 2 weeks, these disturbed sites frequently reached moisture levels low enough to sustain ignition, whereas the undisturbed primary forest never did.

The development of the intitial ignition flame into a fire depends on the presence of adequate fuel to pre-heat adjacent vegetation and to sustain an adequate flame to ignite it. The nature of the vegetation, along with the prevailing, local climatic conditions, will influence this feature of ignition by determining the amount of fuel available, its flammability and its continuity. An upper limit to the standing crop of litter on the soil surface is imposed by the balance between the rate of litter fall and the rate of litter decomposition.

Ignition and spread are both dependent upon oxygen availability and this factor can be determined by the nature of the vegetation at the site. Coarse fuels that are horizontally matted permit little aeration and are therefore difficult to ignite even when dry. In contrast, fine fuels that are well aerated, such as well-dried pine needles suspended in grass tussocks, will ignite readily.

Probability of fire

It is evident from the above discussion that the occurrence of a fire at a given site is dependent upon a whole suite of characteristics being simultaneously appropriate. The occurrence of ignition must coincide with appropriate fuel and approriate climate at the ignition point and with a continuity of fuel between the ignition point and the site. Grimm (1984) pointed out very clearly that some of these component processes have a substantial stochastic element, making it difficult to predict that a fire will occur at a given site at a given time. Macroclimate is temporally variable but spatially fairly constant. Microclimate is variable over both space and time. Abiotic factors such as firebreaks, topography, soil type and size of vegetational units all vary spatially but are relatively constant over time. Biotic factors vary over various time scales; fuel loads increase with time since last fire and flammability varies with changes in the plant community. The probability of a fire igniting must be a product of the probabilities of these individual components. The presence of high fuel loads and the occurrence of so-called 'blow-up' climatic conditions will only ensure fire if an ignition source is available. Thus, in the pre-human past, when lightning was the major ignition source, many 'suitable' fire conditions may have passed without a fire for the lack of a dry lightning storm. Nowadays, however, with the great variety of ignition sources available, fewer suitable days and sites will escape ignition.

Counteracting the patterns of increased fire frequency due to human ignitions, dissection of the landscape with effective firebreaks such as roads or agricultural land must decrease the probability of a given fire spreading from site to site. In the absence of any increase in ignition frequency, these changes would reduce fire frequencies at a given site.

Fire behaviour

Ecological importance

Once flaming combustion has been achieved, relevant questions include: How hot will the fire be? How rapidly will it spread? What will be the other fire characteristics (i.e. ground- or crown-fire, continuous fire front or spot fires, total area burned)? These questions are important because fire intensity and rate of spread are factors which influence the distribution and abundance of organisms. Fire intensity will directly influence scorch height, and therefore determine how much of the plant canopies are consumed, killed or untouched by the fire. The rate of spread of the fire front will determine the 'residence time' for lethal fire temperatures at a given point, a factor that is relevant for both plants and animals. The continuity of the flame front will determine whether animals might escape back through the flames to the relative safety of burnt ground, and the patchiness of the fire will determine whether viable sources of recolonization remain within the fire boundaries. The completeness of combustion will determine the amount of biomass remaining as cover and as a barrier to erosion.

To some extent, the factors that permit ignition will also influence fire behaviour. For example, well-aerated, fine fuels will burn more intensely and spread more rapidly as well as being more likely to ignite in the first place. However, attempting to understand the factors regulating fire behaviour has been a very active and complex area of research for about 30 years, started in Australia by the studies of A. G. McArthur at

*ummary of environmental and biotic factors affecting fire
behaviour*

Factor	Effect
Fuel load	Determines maximum energy available to a fire; Arrangement of fuel can affect aeration (tightly packed fuels), vertical spread (i.e. into canopy) and horizontal spread (patchy ground fuel); Size distribution of fuel can affect likelihood of initial ignition; Chemistry of fuel can increase flammability (i.e. resins & oils), or decrease it (i.e. mineral content).
Overall climate	Determines vegetation productivity and therefore rate of fuel accumulation.
Rainfall & humidity	Increased fuel moisture, combined with high relative humidity decreases likelihood of ignition, rate of combustion & rate of spread.
Wind	Causes drying of fuel; Increases oxygen available for combustion; Pre-heats and ignites fuel in advance of the front, can produce ignition far ahead of front; Wind direction changes can increase fire front.
Topography	Provides variation in local climate (i.e. fuel moisture, relative humidity, interaction with wind); Permits pre-heating & ignition for fires burning uphill; Can provide natural firebreaks; Partially determines distribution of plant communities of different flammabilities.

the Forest Research Institute, Canberra. The following is a much-simplified discussion of the factors affecting fire behaviour, and readers are directed to the following publications for more detailed discussions: McArthur and Cheney (1966), McArthur (1967), Rothermel (1972), Cheney (1981), Vines (1981), Chandler *et al.* (1983) and Johnson (1992). Table 2.2 lists the main categories of factors influencing fire behaviour that are considered in the following discussion.

Factors influencing intensity

Available fuel
The ultimate determinant of fire intensity is the amount of energy stored in the fuel. Fuel load, or total dry weight of fuel per unit of surface area, is

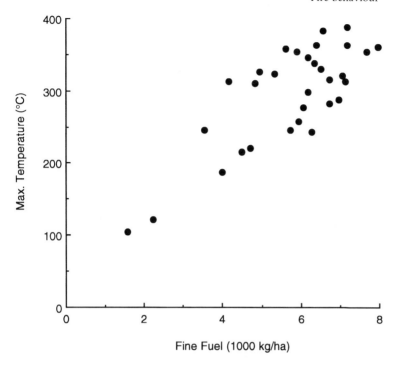

Figure 2.10 Average maximum temperatures of headfires in Texan grassland in relation to the biomass of fine fuel (from Stinson and Wright 1969).

a readily measured indicator of this. There is a strong positive relation-ship between various measures of fuel load and fire intensity, well illustrated in a study of fire intensities in Texan grasslands (Stinson and Wright 1969; Fig. 2.10). Dry weight should really be viewed as *potential fuel*, because few fires actually achieve complete combustion of the above-ground biomass, of any size classes of fuel, in the column above each square metre of soil surface. Thus, a more realistic measurement, termed *available fuel*, is usually made, with fuel size and arrangement (i.e. compactness) taken into account (McArthur and Cheney 1972). It is important not to confuse potential (= total) fuel and available fuel, but in practice it is very difficult to obtain an accurate prediction of how much of the fuel will actually burn. For example, in cool, damp conditions with little wind, a fire burning through dead grasses and leaf litter may not consume, or even scorch, leaves of the shrubs and trees above. Much of the biomass is either out of reach of the fire front or is unavailable due to high moisture content. In this case, an estimate of available fuel would be the total dry-weight of ground-level fuel per unit area, ignoring biomass in the canopy (e.g. estimates by Bentley and Fenner 1958). In contrast, a fire at the same site burning under conditions of extreme fire danger may consume all leaf and twig material and much of the wood. In such a case, available fuel could be considered to be nearly all above-ground biomass.

Using similar reasoning, the same measurement of total fuel made in two different plant communities does not necessarily permit a comparison of available fuel. For example, a given litter load in a longleaf pine–wiregrass community of north Florida would support a fire of greater intensity and more rapid rate of spread than a similar (or even greater) fuel load in an adjacent mesic, hardwood hammock, because of differences in litter aeration and inflammability. Similarly, equivalent litter-layer fuel loads in adjacent wet sclerophyll forest and rainforest in south-eastern Australia will support very different fires.

The vegetation itself can influence how much of the total fuel in the vertical profile can become available under appropriate conditions. Devastating fires rarely occur in natural longleaf pine forests in the south-eastern USA because they are two-layered (tree canopy layer and grass–pine needle layer) and frequent ignitions by lightning keep surface fuel loads at a relatively low level. However, prolonged protection from fire permits woody shrub species to invade this community (Myers 1985). The annual accumulations of longleaf pine needles in the shrub canopies ensure that a future fire will gain access to the pine canopy and cause substantial crown scorch (Chapman 1932). A similar argument forms the basis for one criticism of clear-cutting as a form of forestry. It is reported that the dense sapling regrowth during the early years of regeneration puts more biomass in the category of 'available fuel' than if the forest had not been clear-cut.

One factor commonly affecting fuel loads in fire-prone ecosystems is the time since the last fire, because the rate of accumulation of litter initially exceeds the decomposition rate. These two variables eventually reach equivalence some time after the past fire, producing an equili-brium fuel load of litter that is specific to the plant community and climatic region (Fig. 2.11). However, at some stage after a past fire, fuel load may reach levels considered sufficient to support a high-intensity fire, should long- and short-term climatic conditions and ignition facilitate it. The basis for hazard-reduction burning as a management tool to reduce the likelihood of uncontrollable (i.e. high-intensity) fires is to keep the total fuel load so low that, even in extreme climatic conditions, available fuel can not be sufficient to sustain a high-intensity

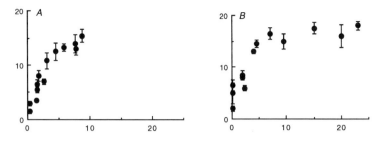

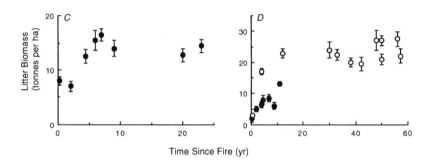

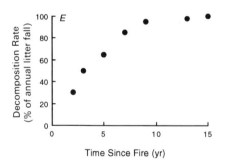

Figure 2.11 Differences in standing crop of leaf litter among sites of different post-fire ages for several eucalypt forest types.

A. *Eucalyptus pilularis/Angophora costata* forest (Fox *et al.* 1979);

B. *E. pauciflora* forest;

C. *E. dives* forest;

D. *E. delegatensis* – pole stand (solid symbols) and older growth stand (open symbols).

E. Rate of accumulation of litter eventually approaches zero as decomposition rate approaches 100% of average annual litter fall for *E. paciflora*. (Data for *B* to *E* from Raison *et al.* 1986.)

fire. The success of burning as a hazard-reduction technique depends upon: (i) a realistic assessment of the proportion of total fuel which will become available in serious fire conditions, and (ii) the ability to burn frequently enough to maintain the levels of available fuel below this threshold.

Moisture and temperature

The climate at the time of the fire, and the preceding climatic conditions, will have a strong effect on fire intensity because the rate of combustion of cold, moist fuels is slower than for hot, dry fuels. Local climate will determine the relative humidity of the air, drying of the fuel prior to the fire, and winds prevailing during the fire. A plant community can exert considerable control over the local climate: a dense, closed canopy reduces evaporation and maintains a high relative humidity and therefore retains moist fuel for some time after each rainfall, and shading reduces fuel temperature.

Chemical factors

As mentioned above, oils and resins in burning fuel increase the heat yield of the reaction because of their greater energy content. Fuels containing high concentrations of these chemicals would be expected to burn intensely. In contrast, relatively high concentrations of mineral elements in wood and leaf material can reduce flammability in some plant species. The role of this factor in regulating fire behaviour at the rainforest – eucalypt forest interface is unknown but would be well worth further investigation. Lindemuth and Davis (1973) suggested that phosphate content of fuel affects the rate of fire spread in Arizona oak chaparral. The potential for modification of fire behaviour by encouraging the development of high-phosphate plant species appears enormous, given the success of using di-ammonium phosphate as a fire retardant in firefighting operations (Foster 1976).

Wind

In the absence of other constraints, a fire generates its own wind. The convection of heated gases upward draws in air from around the burning nucleus and the fire front will spread outward, into this self-generated wind, effectively as a 'back-fire'. The radiation from the front pre-heats the adjacent fuel and thereby prepares it for ignition, and so the fire spreads.

One major effect of wind is to provide more oxygen to the burning fire front. Although this can increase the rate of combustion, the nature

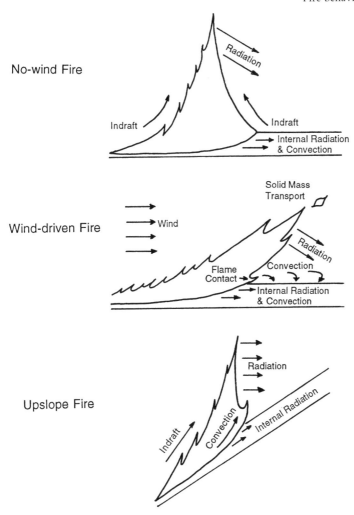

Figure 2.12 The similar influences of wind and topography on fire behaviour are illustrated in these fire profiles. Both slope and wind bring the flames nearer the adjacent, unburnt fuel, so enhancing the pre-heating and increasing the rate of spread (figures from Alexander 1982).

of the vegetation has some control over the magnitude of the effect. Fuels that are densely packed and layered do not permit access to oxygen and therefore constrain combustion. Wind also converts one side of the initial circle of flames from a back-fire to a head-fire. The angle of the flame front, pushed over by the wind, causes rapid pre-heating of the fuel (Fig. 2.12). In the right sort of fuel and with the appropriate winds, this head-fire can produce burning fragments that blow away and set off 'spot' fires in advance of the main front. Striking examples of spotting are provided by the report of the Royal Commission examining bushfires in Victoria in 1939 (Edgell and Brown 1975): 'The speed of the fires was appalling. They leapt from moutain peak to mountain peak, or far on to the lower country, lighting the forests 6 or 7 miles in advance of the main fire.' Other studies report spotting as far ahead of the fire front as 25 km (Foster 1976) and 30 km (Vines 1981).

A change in wind direction may also influence fire behaviour. The effect of a wind change some time after a fire has been burning is to convert the flank of the fire into the front. Komarek (1967) described a grassland fire in Nebraska, USA, which burned with a 5 km front toward the east for 18 km before a 90° wind shift turned it into a fire burning north on an 18 km front.

Topography

The effect of slope on a fire front is similar to the effect of wind. The flames are brought nearer to the ground and therefore pre-heat more of the fuel ahead of the front (Fig. 2.12). The chance coincidence of location of ignition with the topography of the site may also have a strong local effect on fire intensity. A fire ignited on a hill top or ridge is likely to take hold slowly as it burns downhill, whereas a fire ignited in a gully will start more rapidly and gain momentum as it burns uphill.

Another major effect of topography is its interaction with local climate and the patchwork of plant communities. A fire burning up a slope would be expected to burn rapidly and intensely, all else being equal. In reality, however, a large number of additional factors interact to confound such simple relationships. For example, gully vegetation is likely to be somewhat different from hilltop vegetation – perhaps more mesophytic, denser in canopy and therefore less flammable. On a smaller scale, the litter beneath some tree species is likely to differ in flammability, degree of aeration and moisture content, producing local variations in fire intensity (Williamson and Black 1981). A major challenge for modellers of fire behaviour is to take these spatial complexities into account when making predictions about fire characteristics.

Factors affecting spread

Many of the factors affecting intensity of a fire will also determine the rate of spread. For example, fires in dry, windy conditions in a high fuel load will spread rapidly because the rate of combustion is rapid. High fireline intensity will pre-heat adjacent fuel rapidly, and the wind, through its influence over flame length and angle and by causing spotting, will ensure rapid ignition. In addition, other factors are especially important in determining the spread of a fire.

Fuel continuity

Fuel continuity plays an important part in determining whether a fire will spread, especially soon after ignition. Once a fire is burning, fuel continuity will determine the patchiness of the burned vegetation. There are many examples of this. In the mosaic of sedgelands and eucalypt woodlands typical of Hawkesbury Sandstone vegetation near Sydney, Australia (Burrough et al. 1977), fires can burn frequently through the sedgelands but less frequently through woodland patches contained within the sedgelands. Biomass in the sedgelands accumulates rapidly after a fire, and it all becomes available fuel because these plants are flammable when green and the fuel is continuous, much like in a grassland. In contrast, available fuel in the woodland patches consists of leaf litter, dead twigs and woody perennial shrubs. This fuel accumulates more slowly than the sedgeland biomass, and remains patchy for much longer. Fuel continuity, as well as the fuel biomass, is affected by the time since the last fire. In the above example of a sedgeland–woodland mosaic, a long period of fuel accumulation since the last fire in woodland would certainly permit the next fire to be continuous.

Topography

Topographic features can create firebreaks and thereby influence the distribution of fire. Myers (1985) discussed this factor in relation to the distribution of sand-pine (*Pinus clausa*) in Florida. This community type appears to be favoured in situations of infrequent fires, which therefore burn at high intensity in the sand-pine stands, releasing seeds from closed cones. Frequent fire ignitions by lightning and human sources make escape from fire for many years an unlikely occurrence. However, sand-pine stands are frequently associated with topographic features that act as firebreaks. The large 'Ocala Scrub' in central Florida, for example, is bounded on three sides by major river systems (Kalisz 1982). Situations such as these must be common in many regions. For example, Grimm (1984) succinctly described the influence of topography and other firebreaks in determining the fire frequency and hence the distribution of patches of oak woodland in Minnesota prairies. Bergeron and Brisson (1990) found that red pine (*Pinus resinosa*) is restricted to island habitats in

Lake Duparquet, at the northern end of its range in Canada. These islands provide a fire regime that is more similar to those further south, while the mainland in the northern region now suffers large-scale, intense fires that have eliminated this pine.

Plant communities

Plant communities vary in their flammability, for the various reasons discussed above. Less-flammable plant communities can therefore occur interspersed with more-flammable ones, and can act as firebreaks in the same way as seen for topographic features. Another way in which plant community composition can influence fire spread is by the propensity of species within a burning stand to produce firebrands, which are blown ahead of the fire front and produce new ignitions (see Chapter 4 of Chandler *et al.* 1983 vol. 1).

Prediction of fire behaviour

Predictions about the nature of a future fire rest on all the climatic, topographical and vegetational features discussed above and on the assessment of their relative importance in a particular site. This has been perhaps the most active area of fire research, producing a vast array of models that predict fire behaviour. This literature is not reviewed here. Most models attempt to predict the rate of spread and the energy release of a fire burning in a given type of fuel when climatic conditions are varied. In order to obtain a simple model, the numerous climatic and other environmental conditions need to be integrated into a single index. The fire danger meters of McArthur (1966, 1967) are widely used and allow an assessment of the likely severity of a fire if it were to start. Fig. 2.13 provides an example of the relationship between fire behaviour and this fire danger index, given different assessments of available fuel.

Field assessment of fire behaviour

Ecological studies on the effects of fire rarely attempt to quantify fire characteristics, despite the obvious importance of fire behaviour for survival of organisms and functioning of ecosystems (McArthur and Cheney 1966, Cheney 1981, Alexander 1982, van Wilgen 1986). This is quite a serious shortcoming because the applicability of the ecological results obtained in one study to other fires depends upon the fires being equivalent. Although the technology of these measurements need not be

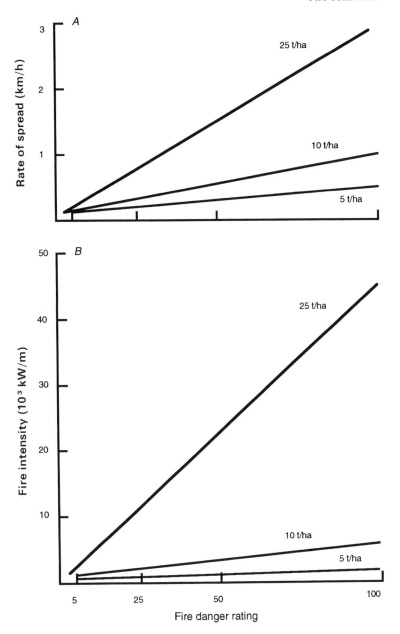

Figure 2.13 Relationship between the 'fire danger rating' of McArthur (1966, 1967) and two features of fire behaviour: rate of spread (*A*) and intensity (*B*) under three fuel loadings of 'available fuel' in Australian eucalypt forest.

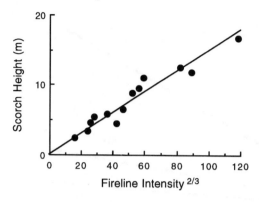

Figure 2.14 Relationship between fireline intensity (effective radiation temperature of the fire front) and height of scorch on the vegetation (from Rothermel and Deeming 1980).

complicated or expensive, some care is needed to ensure that appropriate measurements are made. Measurements of temperature can be made using thermocouples, max.–min. thermometers, or colour-change in temperature-sensitive chalks (see Williamson and Black 1981). Such techniques measure the *maximum* temperature achieved, and therefore convey little information about the overall energy output of the fire. In fact, these measurements are strongly influenced by flame temperature: 'A candle flame has a high "temperature", but since its total heat output is low, a candle can scarcely be considered to provide a "hot fire".' (Vines 1981).

The effective radiation temperature of the fire front (= fireline intensity, Byram 1959) and the total energy released per unit area are perhaps the more important variables to estimate (Rothermel and Deeming 1980). The former measurement is related to the ability of living cells to tolerate heat, and correlates with factors such as scorch height of vegetation (Fig. 2.14). Total energy released per unit area is more closely related to the completeness of combustion and the post–fire state of the site.

Since precise measurements of these components of fire intensity are very difficult to make in the field, they may be estimated by more easily observed parameters such as flame length, rate of spread and residence time (Rothermel and Deeming 1980). Flame length may be estimated visually or from photographs but some attempt should be made to average over, or otherwise account for, the variability inherent in the fire front. Rate of spread of the fire front may be estimated from a sequence of photographs, by direct timing of the front passing known markers or by placing markers in the wake of the front at measured time intervals. Residence time is difficult to estimate because of the indefinite nature of the trailing edge of the fire, but it may be possible under some conditions. This parameter may be more readily estimated by measuring flame depth (Fig. 2.15) and calculating residence time, (t_R), already knowing the rate of spread. Thus:

$$t_R = \frac{D}{R} \text{ (min.)}$$

where: D = flame depth (m); R = rate of spread (m/min.).

Fireline intensity (I) can be calculated as follows:

$$I = 258 \, L^{2.17} \, kW/m$$

where: L = flame length (m),

and energy release (E) (= heat per unit area) is then calculated by:

$$E = 60 \frac{I}{R} \, kJ/m^2$$

where: I = fireline intensity (kW/m); R = rate of spread (m/min.).

Although calculations such as these are valuable, and should play much more of a part in fire ecology than they currently do, they require uniform fuel conditions if they are to be made accurately. This is not a characteristic of natural, field situations. Furthermore, these measurements require an averaging of variables such as flame length and rate of spread. In many cases, it is this very variation that is critical in determining survival of organisms and the magnitude of many other effects. Consequently, a bioassay may be an appropriate way to plot fire intensities after the event. One approach was taken by Koch and Bell (1980), using the degree of scorch sustained by the buried leaves of the monocotyledon *Xanthorrhoea gracilis* (Fig 2.16).

Other bioassays such as this could be sought. Monocotyledons, with their buried meristems and continuous leaf growth provide a visible index of the degree of heat penetration down into the subterranean parts of the leaves of the plant and therefore, by inference, into the soil. For dicotyledons, Scheidlinger and Zedler (1986) and Moreno and Oechel (1989) measured the diameters of the smallest remaining twigs on burned shrubs in chaparral sites in southern California. This measure

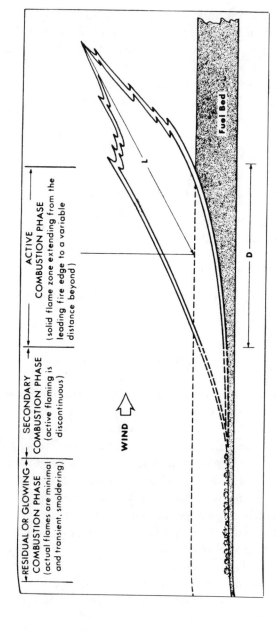

Figure 2.15 Measurements of flame depth (*D*) and flame length (*L*), combined with an estimate of the rate of spread, can be used to calculate fireline intensity (modified from Rothermel and Deeming 1980).

Figure 2.16 The buried leaves of geophytic *Xanthorrhoea* species survive fire and continue growing from the base, thereby carrying evidence of the depth of scorch to the surface soon after a fire.

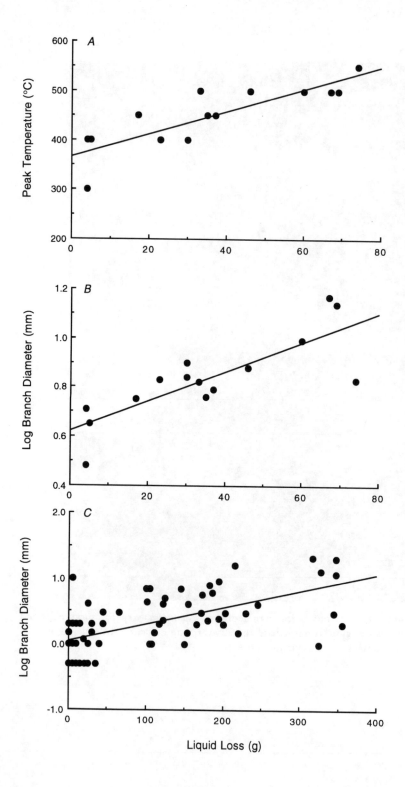

correlated well with another indirect measure of intensity, namely the amount of evaporation of liquid from beer cans placed under the shrubs (Fig. 2.17). Other relevant assays of above-ground fire intensities might be made. One example is the measurement of the degree of opening of closed fruits, which appears to be related to fire intensity in some species, such as pines (Beaufait 1960, Givnish 1981) and *Banksia* (Cowling and Lamont 1985b).

Fire regime

Human perceptions of fire severity are influenced by features related to intensity, such as flame height, rate of spread, extent of the fire front, and magnitude of the smoke pall. It is now well accepted, however, that one fire is not necessarily like another with respect to its impact on the biota. One fire can have a negligible effect on the community within which it occurs, whereas another that is perhaps less 'severe' in human perception can alter community structure markedly. This has been illustrated very clearly by Zedler *et al.* (1983) in a study of the effects on California chaparral vegetation of two fires in consecutive years. A fire in 1979 produced effects well known from other studies: death or pruning-back of some mature plants and stimulation of seed germination, with the expectation that the post-fire vegetation would return to a composition and structure similar to the pre-fire situation. A fire on the same sites in 1980 produced enormous changes in species relative abundances (Fig. 2.18) because of mortality of plants that would normally sprout after fire and also because soil seed reserves, depleted by germination after the 1979 fire, had not been replaced by 1980. Thus, time between consecutive fires, or *fire frequency*, can have a marked impact on vegetation, independent of the fire intensity.

This example illustrates that other features of fire, in addition to *intensity*, may also be important in determining the effects of fire on the biota. These features include the time between fires (*frequency*, as in the

Figure 2.17 *A*. Linear positive relationship between peak fire temperature and liquid loss from 'pyrometer' cans (1 litre cans of water with a small hole perforated in the cap to permit the escape of steam).
B. Correlation between smallest branch diameter remaining on burned shrubs and liquid loss from pyrometer cans in Californian chaparral (Moreno and Oechel 1989).
C. A similar correlation was obtained by Scheidlinger and Zedler (1986) using 375 ml beer cans as containers and beer as the liquid.

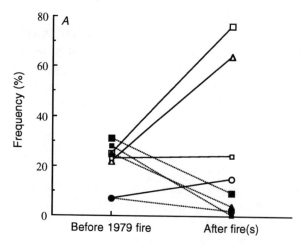

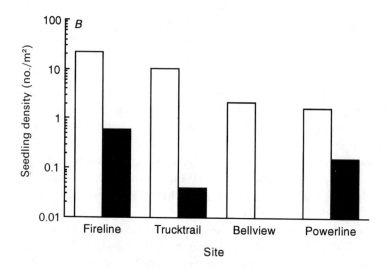

Figure 2.18 Two fires in quick succession can influence plant community structure very rapidly.

A. Frequency of occurrence of *Adenostoma fascicularis* in quadrats sampled before and after fire. At each of four sites (different shaped symbols), paired plots were burned either once (open symbols: only in 1979) or twice (solid symbols: 1979 and 1980). *B*. Densities of *A. Fascicularis* seedlings in these four sites in plots burned only once (open bars) or twice (solid bars). (Data from Zedler *et al.* 1983.)

above example), *season* of burning, the *extent* (or *patchiness*) of a fire and the *type* of fire (i.e. solely above ground versus also consuming the organic layer of soil). These features, collectively, are now well known as constituting the *fire regime* (Gill 1975, 1981a,b).

The term fire regime is becoming widely used in the fire ecology literature but it appears to be gaining two meanings. First, it is used as a description of a particular fire, or of a prescription to be applied to an area. In this sense, it indicates that all the ecologically significant aspects of a fire are being considered. The second, more common use of the term is to summarize the characteristics of the fires that typically occur at a site: e.g. 'What was the pre-European *fire regime* in Australian heathlands?' or 'How did Indians influence the *fire regime* of the California chaparral?' (see Vol. 1, Ch. 9 of Chandler *et al.* 1983).

These are both important concepts. The responses of the organisms in an area to a given fire will certainly depend on *all* the characteristics of the fire. A question such as: 'How will a fire affect ground-dwelling mammals?' requires further refining by asking also: 'In what season; at what intensity; when was the last fire; how extensive or patchy is the fire?' The ecological importance of questions of this type is well illustrated in the examples of the impact of fires on plant populations described in Chapter 5.

The historical fire regime undoubtedly contributed to the evolution of some of the current characteristics of the organisms present and evolution in response to fire as a selective force is explored further in Chapter 3. Understanding evolutionary patterns will require a knowledge of the typical fire history of a vegetation type or region. This is difficult information to obtain, and this field of investigation is in danger of becoming confused through the use of non-standard terminology (Fox and Fox 1987, Romme 1980), especially in relation to fire frequency. Fox and Fox (1987) distinguished between *fire interval* and *fire period* as two components of frequency. Fire interval is defined as the length of time between one fire and the previous fire, fire period is the fire interval averaged over a number of fires (this is referred to as 'average fire interval' by Johnson 1992). The latter is sometimes referred to as the 'fire cycle', although this term is used more often to describe the time interval required for a particular area of a landscape to burn (Johnson 1992). Care should be taken to ensure that the use of these measures, especially fire period, does not obscure variation in fire intervals that can be ecologically very significant.

What factors affect the components of fire regime?

The ultimate regulating factor for any fire regime is climate. Past climate influences fires by determining the characteristics and distribution of the plant communities and current climate determines natural ignitions (i.e. lightning) and subsequent fire behaviour. Many other factors in combination are superimposed on the background of climate to produce even greater variations in the fire regime.

Intensity

The intensity of a fire, once ignited, will be influenced by the range of factors described in previous sections of this chapter: namely climate, topography (slope and aspect), fuel load, fuel type and chemistry, and vertical and horizontal distribution of fuel. The fire history of a site can have a marked effect on the intensity of a fire, via fuel availability. A site that was last burned recently will have been unable to accumulate adequate fuel to support an intense fire. Thus, there is a close link between fire intensity and fire frequency at a site.

Season

The over-riding determinant of fire season will be the climate, because this determines the season of occurrence of natural ignition sources such as lightning. The seasonality of lightning strikes can be measured easily for a given location and related to the timing of plant growth and drying out. Typical fire seasons for naturally occurring fires can be defined for a given region (Fig. 2.19).

Extent

Once ignited, the extent or patchiness of a fire will be influenced by many features of fire behaviour. Heterogeneity of the landscape is the principal one, because natural topographic features such as ridges, gullies and lakes can provide natural fire breaks. Vegetation heterogeneity associated with soils or topography can also be important because some plant communities may act as natural firebreaks.

As mentioned above, a recent fire will make a site less flammable and the next fire more patchy because the fuel load and distribution will have been reduced. The extent and patchiness of a fire will consequently be influenced by the spatial patterns of past fires. Similarly, fires burning 'out-of-season' when the vegetation is still moist and/or climatic conditions are cold and damp, will be of low intensity and also patchy

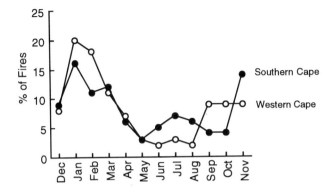

Figure 2.19 Typical fire seasons can be identified for any geographic region by collecting long-term data on numbers of wildfires occurring per month. This figure (modified from Kruger and Bigalke 1984) contrasts the Western Cape (open symbols: most fires in spring and summer) with the Southern Cape (solid symbols: more spread out into winter as well as spring and summer).

(Fig. 2.20). Various aspects of the climate during a fire will also affect its extent and patchiness. Changes in wind strength and direction alter the locations of fire boundaries and factors such as 'spotting' (firebrands being blown ahead of the fire front and starting new fires) are all factors that can produce patchiness in a fire.

Type
The type of fire, i.e. whether it burns the organic soil layers and/or the vegetation and litter, must ultimately be determined by the presence of an organic soil layer. This will depend upon the past vegetation and climate of the site. Adequate moisture and an appropriate pH will be necessary for the development of a peat layer. Ignition and flammability of an organic layer such as peat will depend on extreme, dry conditions. However, once ignited, such fires are likely to burn thoroughly and for a long time.

Frequency
The potential fire frequency at a site will depend principally on two factors; the time required to build up a load of available fuel since the last fire (fuel productivity) and the frequency of ignitions. Superimposed upon this is a variability caused by climate during the 'ignition season' each year. Thus, a high fuel load and lightning strikes will not ensure a fire in a given year. A climate with a marked dry season supporting a

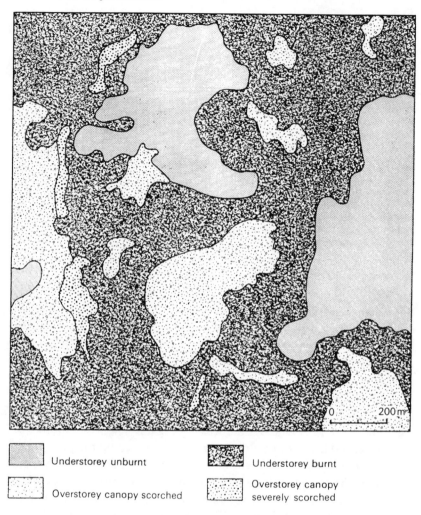

Understorey unburnt

Understorey burnt

Overstorey canopy scorched

Overstorey canopy severely scorched

Figure 2.20 Tracing from an infra-red aerial photograph of a cool-season, prescribed fire in a Western Australian eucalypt forest, illustrating spatial variation in fire intensity (from Christensen and Kimber 1975).

highly productive, flammable vegetation will nevertheless support a low frequency of fire if ignitions are infrequent, i.e. if electrical storms are rare events and there is no human activity. Under these conditions, however, each fire is likely to be widespread and of high intensity because of the great fuel build-up between ignitions.

There are several good examples of the importance of ignition frequency in determining fire frequency, Walker (1981) noted that

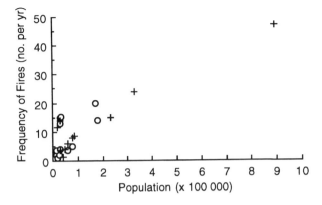

Figure 2.21 The relationship between the frequency of forest fires and the population size of 23 administrative districts in the Hiroshima Prefecture in Japan (data from Takahashi 1982). Open symbols are counties, crosses are cities.

populated areas in Australia, such as around Melbourne, Sydney, Alice Springs and Darwin, have higher fire frequencies than other areas with the same general climate and vegetation type, because of a higher frequency of ignitions due to human activities. Takahashi (1982) found a strong relationship between ignitions of wildfires and local human population density in the Hiroshima Prefecture in Japan (Fig. 2.21). Only 0.7% of forest fires that occurred in Japan between 1946 and 1977 were started by lightning; principal causes were escapes of controlled fires for land management, burning of rubbish, and accidental ignitions (59.7%). Natural fire was historically a very rare feature in this region of Japan, and Takahashi argued that the human-caused change in fire regime has caused a marked shift in vegetation composition. Examples such as these indicate that the proximate control of fire frequency in historic times was ignition source and not regional or local climate or fuel loads.

In contrast to this conclusion, Minnich (1983) has argued that the proximate determinant of fire return times in southern California plant communities, especially chaparral, is fuel build-up, and that natural ignition sources in the past have been frequent enough to ensure a fire whenever fuel loads are sufficient. The corollary of this is that *suppression* of wildfires, including those started by lightning, has produced an increase in the area of vegetation in the region able to sustain fire: when wildfires do occur, they are therefore much larger in area than they would have been in the historic past. Johnson (1992) argued that the

North American boreal forests are similar to the post-suppression California chaparral system as interpreted by Minnich (1983), because long fire intervals permit sufficient fuel build-up to support extensive fires when ignition and uncontrollable spread finally occurs.

Constructing a fire history

The construction of a fire history for a site can make an important contribution to understanding the current biotic characteristics of an area, or to the formulation of a management objective for a particular reserve. In fact, the stimulus for a substantial number of studies of fire histories appears to be a desire to know the historic fire regime on which to base a current fire-management plan.

Many techniques are available for making estimates of an historical fire record (Mutch 1980, Stokes and Dieterich 1980, Johnson 1992). These include: sampling fire scars on trees for evidence of a sequence of fires in the growth rings (e.g. Houston 1973, Dieterich and Swetnam 1984); sampling of sediments in lakes and swamps or peat in bogs for pollen and/or charcoal fragments (e.g. Singh et al. 1981, Cope and Chaloner 1985); sampling lake and reservoir sediments for extreme or unusual runoff events (e.g. Clark and Wasson 1986); using written and oral histories (e.g. Lorimer 1980); and extrapolation from current patterns of weather, fuel build-up and lightning fires. Fire cycles have been estimated, using this range of techniques, for a large number of plant community types in different regions (Table 2.3).

In approporiate situations, the detection of fire scars can provide a great deal of information about past fires, and of the techniques available to construct a fire history, this one has been most frequently used, especially in North America (Stokes and Dieterich 1980). If trees are very long lived, they may contain the record of a substantial 'run' of fires. It must be remembered that the record for a given tree provides a picture just for that tree, not for the stand, catchment, or other vegetation unit as a whole. If sufficient scarred trees are sampled, this picture of historic fire frequency can be expanded to include extent of fires. However, except in some unusual circumstances (Dieterich and Swetnam 1984), this approach cannot provide information on season or intensity of fires, two components of fire regime which are of considerable ecological significance.

Modern remote-sensing technology such as LANDSAT offers many possibilities for precise estimation of some fire characteristics (e.g.

Table 2.3. *Summary of the great range of fire cycles in various ecosystems*

Biome/ecosystem	Location	Fire frequency (yr)
Tundra		
Tundra	Canada, Alaska	500
Alpine tundra	New England	1000+
Boreal forest		
Open *Picea* forest & lichens	Alaska, Yukon	130
Picea glauca forest, flood plains	Alaska, Yukon	200+
Pinus banksiana forest	Northwest Territories	25–100
P. contorta forest	British Columbia	50
Sub-alpine forest		
P. contorta forest	Montana	25–150
	Sierra Nevada, California	100–300
Abies balsamia forest	New England	1000+
Moist temperate forest		
Pinus palustris/Andropogon forest	SE United States	3
Nothofagus forest	Tasmania	300
Eucalyptus forest	Tasmania	100
Grasslands		
Grasslands	Tasmania	10–25
Prairie	Missouri	1
Swamps and marshes	SE United States	30–100
Dry temperate forest		
Mixed conifer forest	California	7–10
Mixed conifer/*Sequoia* forest	California	10–100
Mediterranean-climate vegetation		
Evergreen chaparral	California	20–50
Deciduous chaparral	California	30–100
Semi-arid deserts		
Desert scrub	Arizona	50–100
Pinus & *Juniperus* woodland	W United States	100–300
Tropical vegetation		
Moist evergreen scrub	Florida	20–30
Tropical forest	Equatorial areas	Never?

Source: From Chandler *et al.* (1983).

return-times, areas and season) in the recent past, provided the data become more accessible to ecologists and geographers. Minnich (1983) used a series of LANDSAT images for northern Baja California and southern (USA) California to compare average fire return times in the two areas, which have different human attitudes to fire suppression. Press (1988) used LANDSAT images to assess the extent and season of fires over a 6-year span in northern Australia.

Use of the concept of fire regime

The concept of fire regime is important because it prevents us from treating fire as a single event, from the point of view of the organisms in a site. It is clearly seen as a multi-faceted variable. A given region can be classified, within broad limits, according to the various aspects of fire regime. It is possible to predict, in general terms, the most usual types of fire. Zones of seasonal fire occurrence have been identified for many regions, as described above. For example, Mediterranean-type climates of the world display similar fire regimes, at least in some respects, with fires tending to occur in late summer and early autumn and with fire intervals ranging between 5 and 50 years. Fires may frequently be of high intensity and cover large areas.

There are several major difficulties with the application of the concept of fire regime. First, it can be very difficult to determine the fire history of a particular site from current measures of lightning strikes and fuel loads associated with inferences based on fire-scars, charcoal and anthropological evidence. Consequently, another approach has been to base an inference on the life-history characteristics of the plants dominating the site, or from the relative abundances of species with different life-history characteristics believed to be favoured under different fire regimes (e.g. proportion of obligate seeder and sprouter species). With this approach, an estimate of the fire history is not independent of the characters of the flora. This is a serious problem indeed, if evolutionary interpretations of the effects of fire on the biota are being sought. Second, the characteristics of a given fire are not independent of the occurrence of past fires. Characteristics of the vegetation at a site, which are partially a product of past fires, have a strong influence on the intensity, extent and type of future fires. Third, the stochastic nature of the timing and location of lightning strikes in relation to local climatic conditions at the time and the presence, extent and flammability of fuels, ensure enormous variability in fire regime over time and space within a climatic region. It is therefore impossible to define precisely the past fire regime of a particular site within a region. Furthermore, a single extreme of intensity, season or time between two consecutive fires in the past, such as exemplified by the case mentioned above (Zedler *et al.* 1983), may have long-lasting effects. Local extinction of all species with a life-history sensitive to the extreme event would lead to an incorrect inference about the *average*, past fire regime of the site.

An added complication is the fact that human activities have altered various aspects of fire regime. Fire frequencies have been reduced in forest regions due to fire-suppression policies; they have been increased due to the greater frequency of ignitions associated with centres of population and altered intensities have followed these changes in frequency; fire extents have been reduced because roads, towns and agricultural clearing all act as firebreaks in addition to natural topographic features; and fire season has been altered with the introduction of cool-season, hazard-reduction burning.

Application of the concept
An important application of the concept of fire regime relates to estimating the effects of altered fire regimes, caused by human activities, on the biota. Unplanned fires frequently burn into National Parks, Nature Reserves and specific management areas (e.g. water catchment areas). We might expect the organisms currently living in an area to have the capacity to tolerate a continuation of the past, 'regional' fire regime. As a corollary, we might also expect severe departures from a fire regime to influence the biota to some degree. Management of natural communities in many regions now includes the use of prescribed, controlled fires to reduce the intensity of any future wildfire. An understanding of past fire regimes is necessary here, because serious departures from other aspects of the past fire regime (i.e. in season, frequency, extent) may be expected, as a first approximation, to have a deleterious effect on the biota.

Outstanding questions

The breadth of material covered in this chapter is substantial and much of it provides a general background to the abiotic features of fire. It is clear that a lot is known of the physical and chemical effects of fire and much of it is relevant to an understanding of fire ecology. Perhaps the most significant general message is that physical characteristics of fire are far too rarely estimated or published in relation to ecological studies. More attention must be paid to this in the future, by ecologists, especially because adequate replication of fires in experimental studies is so difficult. Comparisons of fires occurring at different times or in different locations can become useful only if the fires are adequately described.

There are many gaps in our knowledge of ecologically relevant fire characteristics, and obvious gaps are listed here as an indication of some profitable directions for future study.

1. Fire histories based on fire scars recorded in tree rings are needed in regions in addition to North America. Also, inferences of fire histories based on a range of approaches are needed.
2. Because of the potential ability of extreme fire events (e.g. extreme intensity, extremely short – or long – return times, extremely out of season) to alter the biota greatly, more attention needs to be placed on quantifying this aspect of the fire history in an area. How much do the extremes of fire intervals in the past vary from the fire period?
3. Related to this, spatial variation in topography and local climate may have had a predictable influence on the spatial patterns of past fires. This spatial heterogeneity in fire likelihood can have considerable ecological significance (e.g. refugia for fire-sensitive species), but is obscured when the average fire return time is calculated simply from an estimate of the area burned per year divided by the total area of the region.
4. Data on intensity–height and depth–height profiles are needed for a variety of vegetation types, especially for the canopy – which contains meristems and seeds – and for the soil zones containing roots and seeds.
5. Quantification of the spatial variation in fire intensity should be attempted in many communities.
6. Design of simple-to-use and repeatable ways of estimating fire characteristics that are of ecological relevance (e.g. pruning-by-fire of Moreno and Oechel 1989; beer cans of Scheidlinger and Zedler 1986) would be of great advantage to future studies of plant, animal and community responses to fire.
7. Measurement of post-fire physical conditions, especially soil and surface temperature regimes and wind speeds, are lacking in most fire studies yet may be of even more significant to the biota than the fire event itself.

3 · Survival of individual organisms

The survival of individual organisms through a fire will be determined by various life-history, anatomical, physiological and behavioural characteristics. Population and community changes in response to a particular fire regime (see Chapters 4–7) will depend, to a large extent, on these characteristics of the individual organisms.

First a warning about naively classifying organisms according to characteristics observed to allow them to persist in spite of fire. Such characteristics have been referred to as 'fire-adaptive traits' (Gill 1981b, Keeley 1981). There are several arguments against the use of such a term. The word 'adaptation' gives the erroneous impression of an inevitably close fit with present or future conditions (see discussion in Begon *et al.* 1986, pp. 5 and 6). The term also implies that fire was the sole selective force producing the character (Williams 1966, Frost 1984, Whelan 1985). Finally, the responses of a given species depend on the characteristics of the particular fire under observation. As explained in Chapter 2, fire is not a constant phenomenon but comprises several variable characteristics including intensity, season, frequency and extent. A species favoured in one fire and exhibiting strong recruitment may well suffer mortality and failed recruitment after another fire of different characteristics. Thus, a trait considered to be adaptive in the first fire may well be considered maladaptive in the second.

Despite these potential problems, a classification of the characteristics of organisms that permit survival of particular fires, if it is interpreted carefully and related to well-described fire characteristics, can point to the possible roles various features of a fire regime have had in moulding organisms during the evolutionary past and can also allow predictions to be made about the survival and reproduction of organisms in a particular fire. A discussion of evolutionary responses to fire appears at the end of this chapter.

Most treatments of the effects of fire on individual organisms start by describing spectacular examples of those species in fire-prone ecosystems that are able to tolerate or escape the most severe fires. A different approach is taken here, by building on the fundamental question of what the effects of fire are on living tissues. The answer to this describes the potential threat to survival posed by a fire. Logically following from this is the range of solutions displayed by different organisms in different environments. Survival of fire has two components: (i) the survival of the direct effects of fire, experienced during the passage of flames, and (ii) tolerance of the changed post-fire conditions. It is important to recognize this dichotomy, because many organisms find ways of tolerating the passage of a fire, only to succumb to stresses imposed by post-fire conditions.

It can be argued that plants and animals may have evolved fundamentally different solutions to the threat of mortality by fire, by virtue of the relative immobility of plants compared with most animals and by virtue of the fact that plants are more 'modular' than animals and can therefore tolerate serious injury to some of their component parts. It is logical to treat these groups separately and plants will be examined first.

Tolerance of plants to fire

Cell and tissue death

The temperature experienced by plant cells is an ultimate cause of tissue death in a fire. Studies on the physiological responses to high temperatures have been conducted mostly in relation to continuously applied, ambient heat (Alexandrov 1964, Levitt 1972). Under these conditions, damage caused by high temperature can be either indirect, following metabolic changes, or direct, caused by protein denaturation, altered lipid mobility or chemical decomposition. There is little published information about the effects of the relatively short pulse of intense heat experienced by plant tissues during a fire. At one end of a scale of severity, complete combustion obviously causes cell and tissue mortality. At the other end, a brief and slight elevation in temperature may cause only a temporary disruption of some biochemical pathways (Fig. 3.1).

For a given temperature applied to a plant tissue, variation in cell mortality results principally from (i) the length of time the heat is applied, and (ii) the state of the cells, especially whether they are hydrated and metabolically active. Levitt (1972) reviewed the relationship between the temperature applied and the duration of application that can cause death (Fig. 3.2). This relationship clearly varies among

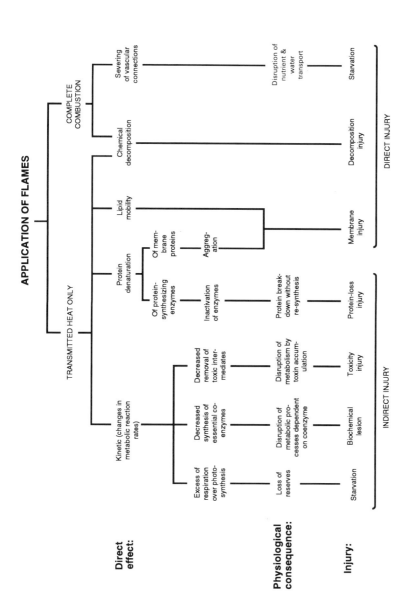

Figure 3.1 Range of pathways through which the application of heat to plants causes direct and indirect damage (modified from Levitt 1972).

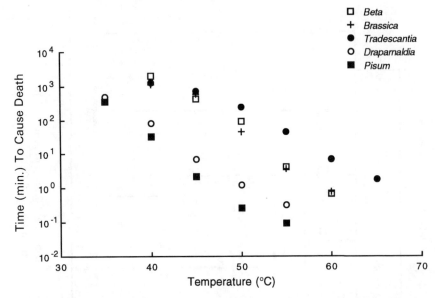

Figure 3.2 Semi-logarithmic relationship between the temperature experienced by a plant tissue and the time taken to cause death of cells (from Levitt 1972). The precise nature of the relationship varies among the five plant species tested, with *Tradescantia* tolerating even the highest temperature for over 1 min. *Pisum*, in contrast, tolerated only 45 °C for this length of time.

species, but for each it appears to follow a semi-logarithmic relationship (*lethal temperature* $= a - b$ log *heating time*).

Resting plant tissues that are in a dehydrated state can tolerate a much more severe heat treatment than can tissues that are metabolically active and fully hydrated. Striking demonstrations of this are experiments with seeds. In an experiment by Schneider-Orelli (cited by Levitt 1972), intact, hard-coated seeds survived autoclaving at 120 °C for 0.5 h. Similar seeds were killed by boiling for only 10 min. if the seed coats had been scarified first, permitting hydration. This relationship between degree of hydration and susceptibility to high temperature apparently holds for other plant tissues as well as for seeds (Levitt 1972).

The above discussion has concentrated on the mortality of cells and tissues actually exposed to a particular temperature. The likelihood of death of a whole plant will depend upon both the extent of injury to its component parts and which tissues are affected by the heat. Some parts of a plant are more important than others for its continued survival after fire. The cambium is critical tissue for stem and canopy survival because of the vascular bundles it contains. Rundel (1973), for example, reported a strong correlation between crown dieback and fire scars at the base of the trunks of giant sequoia (*Sequoiadendron giganteum*). Meristematic tissue is important because the production of new leaves and flowers after injury by fire depends on the survival of these buds. Seeds are critical too, because they represent the genetic future of a plant. Seeds also represent, for many species, the only opportunity for mobility during the life cycle.

There are two ways for parts of a plant to tolerate exposure to fire. One is for cells comprising the critical tissues to withstand the biochemical degradation resulting from heat; i.e. to have a higher lethal temperature. The other is the protection of these critical tissues by preventing them heating up to the lethal temperature. The degree of hydration of critical tissues and their state of metabolic activity at the time of a fire will influence their survival. However, the capacity of individual cells to withstand a constant exposure to heat is not considered to vary significantly either between plant species or between tissues within a plant. In general terms, the thermal death point for cells of typical, mesophytic plants is considered to lie between 50 and 55 °C (Hare 1961). The explanation for spectacular feats of survival by plants exposed to intense fires must therefore lie in the protection of their critical tissues from excessive heat.

Protection of critical tissues

There are several ways in which those tissues important for post-fire recovery of a plant are protected from lethal temperatures during the passage of a fire front. First, cambium and meristematic cells can be shielded from radiant heat by an insulating bark. Second, above-ground vegetative parts can be sacrificed and subterranean stems and meristematic cells insulated by the overlying soil. Third, sensitive tissues can be borne at such a height above ground that they would be unlikely to suffer intense heat. Similarly, seeds can be protected by insulated fruits, by burial in the ground, or perhaps simply by the height at which they are borne in the canopy.

Bark

From the 1930s to the 1950s much effort was directed at studies of the insulating characteristics of bark of trees in forests of south-eastern North America (Spalt and Reifsnyder 1962), perhaps stimulated by the increasing pressure to manage conifer stands using prescription burning.

Numerous studies conducted during those years indicate that the time taken for cambial cells to reach a lethal temperature, when heat is applied to a tree trunk, is a function of both bark thickness and thermal properties of the bark. These factors vary with plant size and also among species. The relationships are illustrated in a comparative study by Hare (1965), in which a flame from a propane torch was applied to the bark and the temperature of the underlying cambium measured with a thermocouple inserted beneath the bark. The time taken for the cambium to reach the assumed lethal temperature (60 °C) was exponentially related to bark thickness for all species of trees tested (Fig. 3.3A) and the positive correlation that exists between tree diameter and bark thickness produced a close relationship between diameter at breast height and time taken for its cambium to reach the lethal temperature (Fig. 3.3B). Thus, small trees of a species are more susceptible than large ones, because of the allometric relationship between bark thickness and stem diameter. Fig. 3.3 also illustrates a substantial variability among species, for any particular bark thickness. One reason for this variability is that different types of bark have different thermal properties, principally because of differences in moisture and air content (see review by Spalt and Reifsnyder 1962).

An important feature of the studies described above is that the time taken for the cambium to reach a particular temperature (i.e. 60 °C) depends upon the starting temperature (Hare 1961), so a cambial temperature of 60 °C is reached more rapidly when ambient temperature is high in summer than during cold winter conditions. This effect of ambient temperature may partially explain the greater resistance of some tree species to winter fires (Cary 1932, Hodgkins 1958), but there are other possible reasons too, including the physiological state of the tree at the time of burning and the confounding factors of variable fire intensity and behaviour.

Gill and Ashton (1968) obtained essentially similar results to these North American studies, in a comparison of the insulating effectiveness of the bark of three co-occurring *Eucalyptus* species in Australia. For each species, increasing bark thickness delayed the increase in cambial temperature. However, the exact nature of this relationship differed between the species (Fig. 3.4). One reason for the differences appeared to be differing flammability of the outer layer of bark. There are several classes of bark types in the Australian eucalypts: ranging from smooth, white bark to rough, fibrous 'stringybark'. The dry outer bark of the stringybark species (*E. obliqua*) ignited almost immediately on appli-

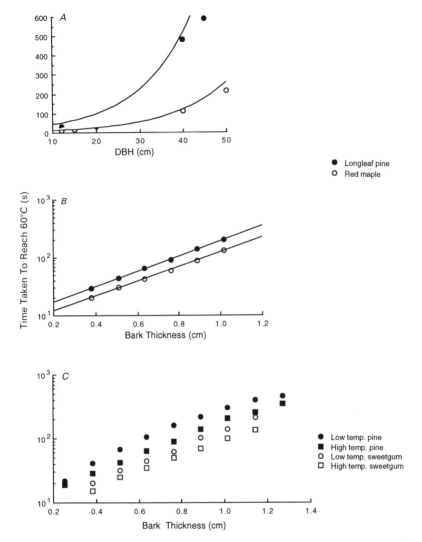

Figure 3.3   The rate of heating of cambium when fire is applied to a tree trunk varies among species. For a given species, heating is related in various ways to the nature of the bark (data from Hare 1965).

A. The rate of increase of temperature of the cambium increases exponentially with tree diameter in longleaf pine and red maple. DBH = diameter at breast height.

B. This effect of tree diameter can be explained by the relationship between bark thickness and diameter. For a given thickness, longleaf pine bark retards the rate of increase in cambial temperature better than the hardwood bark.

C. The time taken for cambium to reach a presumed lethal temperature (60 °C) is related to the starting ambient temperature. Thus, with a low ambient temperature (0.6 °C – circles), cambium takes longer to reach 60 °C than with an ambient temperature of 16 °C (squares) for both pine (solid symbols) and sweetgum (open symbols).

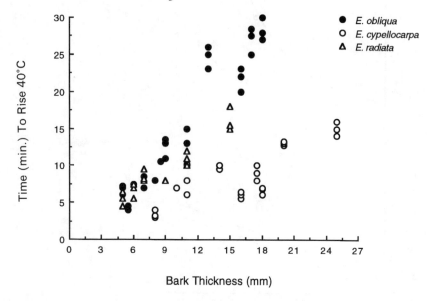

Figure 3.4 The influence of bark as an insulator of the cambium in Australian *Eucalyptus* species is essentially similar to that for the North American species (see Fig. 3.3): with thicker bark, the cambium takes longer to rise through 40 °C (data from Gill and Ashton 1968). However, the quantitative nature of the relationship varies among species. For a given bark thickness, *E. obliqua* (solid circles) takes longer to heat up than either *E. cypellocarpa* (open circles) or *E. radiata* (triangles).

cation of the heat source, whereas bark of the white, smooth-barked species (*E. cypellocarpa*) started charring only some time after the initial application of heat, and did not ignite. In addition to these differences in bark flammability, energy reflectance from the bark surface varied among species, being greatest in the smooth-bark species and least in the stringybark species.

Gill and Ashton's study illustrates another feature of bark, namely its thermal 'inertia' – a term that refers to the retention of heat in the cambial layer after the source is removed (Fig. 3.5). This is relevant because of the inverse relationship, described above, between temperature attained and the time required to cause cell death. Thermal inertia of bark is also significant from a practical point of view. Measurement of temperature during a fire is frequently achieved using thermal chalks, which melt as soon as they reach a given temperature. Because cell death is influenced both by the maximum temperature experienced and also by the duration of high temperature, the peak temperature measured using techniques like this would be expected to be only loosely correlated with cell and tissue death.

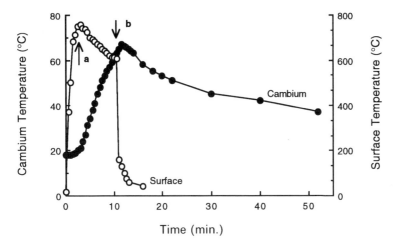

Figure 3.5 Thermal inertia of bark determines the length of time a particular temperature is maintained at the cambium. In this example the outer bark of *Eucalyptus obliqua* ignited upon heating and completed flaming at time 'a', when surface temperature (open symbols) peaked. The heat source was removed at time 'b'. Cambial temperature (solid symbols) peaked soon after 'b' but took a long time to return to a low temperature (data from Gill and Ashton 1968). (Note that the scale for cambial temperature is expanded to 10 times that of surface temperature.)

The above examples illustrate the way in which fire *intensity* can cause tissue death, namely by directing heat at the bark for a sufficient time. Fire intensity would be increased by a heavy fuel load, as influenced by fire frequency and season of burning. As explained above, the tolerance of the underlying tissue of the peak temperature it experiences will also be influenced by *season*, through the effects of cell hydration and metabolic activity and the *frequency* of fires will influence plant survival if the recovery of a given bark thickness after one fire is not complete before the next fire occurs. Gill (1980) showed that the recovery process may take more than 7 years for a smooth-barked *Eucalyptus* species (*E. dalrympleana*). In a given fire, the head-fire and back-fire seem likely to have different effects, because the back-fire subjects the trunk to elevated temperature for a longer duration than does a head-fire (Fahnestock and Hare 1964; Fig 3.6).

It is common to see fire scarring on the uphill side of tree trunks. Several factors appear to act in combination to cause this phenomenon. In sloping country, there is frequently an accumulation of leaf litter and twigs against the uphill side of tree trunks. This would result in a locally higher fire intensity and duration on the uphill side of the trees.

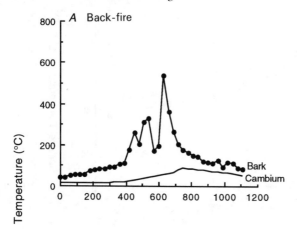

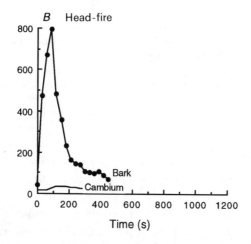

Time (s)

Figure 3.6 A fire burning into the wind (*A*) causes elevated temperatures at the bark (solid symbols) and the cambium for some time, whereas a head-fire (*B*) may produce greater peak temperatures for a much shorter duration (modified from Fahnestock and Hare 1964).

However, even without this asymmetrical distribution of fuel, fire can be concentrated on the uphill side, as demonstrated by Gill (1974) in an elegant series of laboratory simulations (Fig. 3.6). This occurs because of wind eddies, which are pronounced in a fire moving up-hill.

Vegetative insulation

Elementary ecology texts identify one fundamental difference between grasses and dicotyledons which explains their different responses to grazing. Grasses bear their meristems at the leaf bases while dicots have them exposed and constantly elevated as the plant grows. This character-istic of grasses also protects many of them, especially clump and tussock grasses, from fire. As we have seen, much of the heat of a fire is directed upward (Packham 1970). Moreover, dense-packed stems and leaves in a grass clump serve to insulate the meristems, buried deep inside.

Even in arborescent monocotyledons and in dicotyledons, in which meristems are more exposed to heat than in grasses, particular leaf arrangements can insulate otherwise susceptible buds from the peak of heat in a fire (Gill 1981c). The genus *Xanthorrhoea* in Australia provides an excellent example of this (Fig. 3.7A; Gill and Ingwersen 1976). The rigid leaf bases of both this genus, and the cycad *Macrozamia*, provide such good protection from heat that many invertebrates survive fire by burrowing deep into their crowns (Whelan *et al.* 1980). Vogl (1969) described a similar growth form and insulative effect in the genus *Pandanus* in Hawaii. Saplings of the longleaf pine (*Pinus palustris*) in North America, have needles densely packed around apical buds (Fig. 3.7B), and the insulative effect of this structure has long been the subject of much interest and experiment (Andrews 1917, Wahlenberg 1946), where tissue paper wrapped around the apical bud remains unscorched in a fire that is of sufficient intensity to burn needles from the stems and to scorch leaves on canopy trees.

Kruger and Bigalke (1984) suggested that the architecture of the whole canopy of some plant species may permit tolerance of fire by deflecting heat away from the apical buds. Of course, stems must also be protected by insulating bark for this to be effective in survival of the plant. This mechanism appears to hold for three species of Proteaceae in southern African fynbos, *Protea laurifolia, Leucadendron argenteum* and *Leucospermum conocarpodendron*. In addition, Kruger and Bigalke identi-fied two species, *Protea effusa* and *Leucadendron glaberrimum*, in which apical buds appear to be insulated by the physical structure of the dense, matted foliage of the canopy. This novel suggestion would perhaps be worthy of direct investigation in species in other fire-prone ecosystems.

Roots and underground stems
Bark is not the only line of protection against high temperature in shrubs and trees. Many plants, especially shrubs, sacrifice meristematic tissues in the canopy but nevertheless tolerate fire by sprouting from previously suppressed underground buds, either in buried stems or in roots. Soil is a good insulator (Priestley 1959) and tissues deeper than 5 cm rarely experience significant elevation in temperature (see Chapter 2).

The insulating characteristics of soil are exploited particularly well by

A

B

Figure 3.7 The potential for insulation of apical meristems is illustrated in *Xanthorrhoea australis* (*A*) in Australia and *Pinus palustris* (longleaf pine) (*B*) in the south-eastern USA (etching from Mohr 1896).

those plant species with lignotubers (or burls; James 1984). This specialized root-crown structure is prevalent among shrubs and trees of Mediterranean regions and among Australian eucalypts. Bamber and Mulette (1978) showed that the anatomy of the lignotuber of *Eucalyptus gummifera* was qualitatively similar to stem tissues. However, lignotuber wood contains a higher proportion of storage tissue (axial and radial parenchyma cells), thereby providing a greater potential for storage of starch reserves. The lignotuber therefore fulfills dual functions of being a source of protected buds and of storing the energy required to support sprouting.

The ability of a plant to sprout from its lignotuber after a fire of a particular peak intensity and duration will depend upon the depth of the lignotuber beneath the soil surface, and perhaps upon the size of the lignotuber. For example, differential survival of two mesquite species in North American semi-desert shrubland is apparently related to the depth of the root crown that contains the buds (Wright 1980).

Fire frequency will also be of importance because of the time required for a plant to replace the energy stores withdrawn from the lignotuber during sprouting and the ability to continue replenishing the 'bank' of buds in the lignotuber. The capacity for repeated recovery after frequent fires appears to be highly variable among species. Zammit (1988) found that four consecutive bouts of clipping shoots of *Banksia oblongifolia* were all that was required to prevent further sprouting. In contrast, Chattaway (1958) found that *Eucalyptus* seedlings could be clipped of all leaves for up to 26 times before finally succumbing.

Season of burning and plant physiological status have also been shown to interact to influence sprouting after defoliation. Cremer (1973) found that for some plant species in Australian eucalypt forests, recovery after defoliation was poor after periods of rapid shoot growth and high after periods of quiescence. Cremer related this to accumulation and storage of starch in both roots and stems: in periods of rapid growth, starch is directed to growing shoots and little is available to support sprouting if defoliation occurs at such a time.

Height above ground

The decline in peak temperature with height, described in Chapter 2, indicates that critical tissues such as seeds and buds may be protected from heat damage simply by height above the flames. The longleaf pine (*Pinus palustris*) community of south-eastern North America (Fig. 3.8) is a good illustration of this. Grass fires on the floor of this open forest rarely

A

B

Figure 3.8 *A*. Two-layered longleaf pine (*Pinus palustris*) forest with tall canopy and low ground layer of herbs, shrubs and grasses. In this type of forest, a ground fire is unlikely to carry into the canopy under most climatic conditions.
B. In contrast, a mid-storey can collect fallen pine needles and provide a 'ladder' carrying fire into the canopy.

cause canopy damage and tree death, but saplings of longleaf pine are susceptible to fire (Wahlenberg 1946, Bruce 1954). The rapid growth in height of longleaf pine saplings has been interpreted as an evolved response to fires, raising the apical meristem out of the 'danger zone' in just a few years. Mature trees can sustain canopy damage in a fire under certain circumstances. When there is a mid-storey, pine needles are caught in the foliage and a fire may carry from the grass layer into the mid-storey layer. Thus, the fire is carried closer to the canopy and canopy scorch and tree death can follow.

This phenomenon of fuel draped in branches in the mid-storey carrying a fire into the canopy has also been mentioned in Australian eucalypt forests (Glasby et al. 1988). This example also introduces the potential importance of season of burning for plant survival, because the annual decortication of bark contributes substantially to the fuel accumulation both on the ground and in the canopy. Fires occurring when this accumulation of fuel is at its peak would be more likely to carry into the tree canopies.

The protective value of height can be relevant to a plant only if the cambium is protected in the stem. Survival of canopy meristems is of little value if fire breaks the vascular connections to the roots (see Rundel 1973). Longleaf pine bark clearly protects the cambium from damage in most fires. Fahnestock and Hare (1964) showed that, although the temperature at the bark surface 30 cm above the ground peaked well above 500 °C, the underlying cambium rarely reached lethal temperatures (see Fig. 3.3). However, damage to the bark close to the ground can negate its protective, insulating effects. The turpentine industry in the south-eastern USA, which used longleaf pine as a principal resource, produced scars on many trees, which now ignite even in relatively low-intensity, ground-fires, causing severe damage or death to the tree. Similar damage to the bark within the reach of ground-fires can also result if tree trunks grow close together or with low forks, causing higher peak temperatures and greater durations on both trunks.

Glasby et al. (1988) related the survival of Eucalyptus oreades trees and saplings to the 'skirt' of corky bark that develops around the trunks of trees older than about 20 years. Low-intensity, prescription fires kill a substantial proportion of trees less than 15 cm diameter (at breast height), apparently because it is only the larger trees that have developed a cork layer sufficiently high up the trunk to be effective in insulating the cambium (Fig. 3.9)

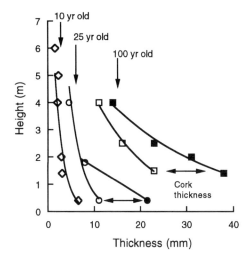

Figure 3.9 The role of bark in insulating tree trunks is related to the height of fire temperature and the height of the skirt of bark. In *Eucalyptus oreades*, both live-bark thickness (open symbols) and thickness of dead, corky bark decline with height up the trunk (see Glasby *et al*. 1988). Bark thickness at any given height is related to the age of the tree.

Flowering

Flowering is one stage of plant reproduction that is particularly suscept-ible to fire, even if the established plant is able to sprout and even if seeds are protected in fruits or in the soil. The relationship of the season of burning to the plant's flowering time can be important, because fire too early would kill flowering buds or developing flowers and thus kill a full year's worth of potential seeds. Moreover, sprouting plants may take some years to recover sufficiently to support flowering again. Situations in which this phenomenon may be expected to be most severe in terms of annual seed production are: (i) when the current year of flowering contributes a high proportion of the accumulated, dormant seed bank (i.e. where depletion of a stored seed bank is initially rapid); and (ii) where a mast flowering year occurs only infrequently and happens to be preceded by the fire. In the latter case, failure of reproduction could be magnified if the burned plant failed to recover from fire quickly enough to flower by the following mast year.

Variation in annual flowering and the dynamics of seed storage in the genus *Banksia* provide an example of the possible impact of timing of fire on reproduction, both in a given year and at a particular season within a year (Fig. 3.10). There are three interacting factors. First, annual flowering may vary markedly from year to year (Copland and Whelan 1989). Second, a fire early in the season may eliminate most or all of the seed production destined for that year. Third, for bradysporous species (in which seed is retained for some time in fruits in the canopy), the cohort of fruits supporting the majority of viable seeds may be those from the most recent year of flowering. Declining viability and pre-dispersal seed predation can cause high seed losses after the first year (Cowling *et al.* 1987, Lamont and Barker 1988; cf. Lamont *et al.* 1991b). Certain combinations of the above three factors could lead to a significant reduction in reproductive success because of fire's effect on flowering (Jordaan 1949, 1965, Midgley 1989). For example, a fire in autumn 1985 (27 months in Fig. 3.10A) would have eliminated a potentially good year for seed production in *Banksia ericifolia* (Copland and Whelan 1989), leaving only the fruits from the previous, poor flowering year (18 months) to disperse a high proportion of viable seeds after the fire – though a small number in total. Many seeds from the previous, good flowering (6 months) may have been eaten or otherwise lost viability. In contrast, a fire at 36 months, after fruit maturation of the prolific flowering, would release large numbers of seeds still with a high viability. The poorer flowering (18 months), by now with reduced seed viability, would have contributed relatively little to the overall seed release in any case.

The above scenarios are predictions based on circumstantial evidence and logical expectations, based on an assumption that bradyspory need not provide a continuing accrual of viable seeds in a seed bank. Although there is some evidence in support of this assumption, it is not always the case. Lamont *et al.* (1991a) showed that seed losses in *Banksia cuneata* are not substantial soon after fruit set. It will be important to examine the dynamics of the stored seed bank in a number of species and populations (see Pannell and Myerscough 1993).

Seeds
The seed is a stage of the plant life-cycle with a built-in advantage in a fire. Seeds contain cells in a dormant and often dehydrated state; characteristics which both, as mentioned above, confer some tolerance of high temperature. Even seeds of domesticated species such as pea, sunflower and wheat were able to tolerate dry heat between 70 and 90 °C applied for 4 h in an experiment by Beadle (1940). Beadle also found that

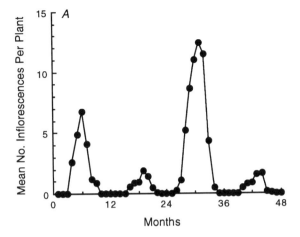

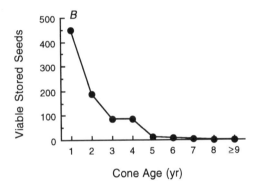

Figure 3.10   The season in which a fire occurs can have an impact on plant reproduction because of underlying variability in flowering intensity and seed storage.

A. Flowering in *Banksia ericifolia* in southeastern Australia can vary greatly from year to year, producing variation in the numbers of seeds deposited annually in the canopy-borne seed bank. Mean number of inflorescences per plant, from a sample of 30 plants, shows marked year-to-year variation with the third year (1985) producing three times as many as the years either side (modified from Copland and Whelan 1989). A fire early in the flowering season of a good flowering year could therefore cause a significant reduction from the potential seed bank.

B. This is especially true when there is a rapid decline in seed viability of stored seeds, such as in *Banksia burdettii* illustrated here (data from Lamont and Barker 1988).

seeds of some native plant species in the fire-prone vegetation of the Sydney region in eastern Australia were even more resistant to this dry heat treatment than the crop seeds. Beaufait (1960) described high tolerance of individual seeds of jack pine (*Pinus banksiana*) to high temperature, with no significant loss in viability after exposure to 370 °C for 15 s or less.

Is the degree of tolerance described above adequate for survival of fire? Temperatures experienced by seeds exposed to the heat of a fire would usually exceed 110 °C. Individual seeds are small and therefore heat up to maximum temperature rapidly. Thus, seeds lying among the leaf litter when a fire occurs are almost certain to be killed (Floyd 1966). To tolerate fire, seeds must be protected from direct heat. The two predominant means of protection are burial in soil and enclosure within fruits in the plant's canopy.

The decline in the magnitude of temperature increase with depth in the soil (see Chapter 2) means that burial provides a potential refuge for seeds, as evidenced by worldwide observations of post-fire germination in burned sites. Cheplick and Quinn (1988) described a plant species that produces its seeds about 3.5 cm underground, well below the depth at which soil temperature would become lethal to seeds in a fire.

Gill (1981b) pointed out that the relationship between seed survival and germination, depth of burial, fire intensity and type, and soil moisture must be very complex. It can be broken into various components. First, mortality of seeds may be expected near the soil surface: just how near will be determined by the intensity and type of fire burning across the surface and by the tolerance of the seeds to heat. Second, although deeper would appear to be better, there is a trade-off because successful germination is impossible from great depth. Auld (1986b) showed that viable, heat-treated seeds of *Acacia suaveolens* germinated readily from depths down to 8 cm in glasshouse trials: below 8 cm there was a steady decline in the proportion germinating, to nil at 14 cm. Putting these two effects of depth of burial together should produce a generalized relationship for depth and germination (Fig. 3.11*A*). Seeds close to the surface would be killed by the fire; seeds at some intermediate

Figure 3.11 Schematic model of seed survival and germination in relation to depth of burial and fire intensity for seeds with no innate dormancy, i.e. hard seed coat (modified from Martin and Cushwa 1966).
A. Seeds close to the soil surface would be killed by the heat of the fire (survival curve – solid line), seeds buried too deeply would remain in enforced dormancy (dormancy curve – dashed line) and successful germination would be the balance of the opposing effects of heat and depth.

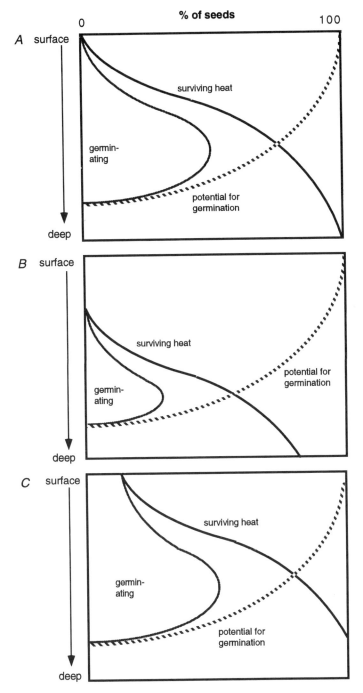

B. With *increasing* intensity, the seed mortality curve would move downwards, compressing the zone of successful germination against the dormancy curve.
C. With *decreasing* intensity, the zone of successful germination will expand upwards with the seed survival curve.

depth could germinate successfully; seeds buried too deeply would remain in a state of enforced dormancy.

With this schematic model, one can predict the effect of increasing fire intensity and duration of heating at the soil surface, such as occurs in a post-logging slash burn. The curve representing seed mortality would move downward through the depth profile. For viable seeds, the relationship between germination and depth would not change, however, so the proportion of the seed bank capable of successful germination would decline, reflected by compression of the successful germination area (Fig. 3.11B). Decreasing the fire intensity would expand the zone of successful germination both through raising of the seed survival curve and also because a lower-intensity fire is likely to be patchier, permitting more of the surface area to escape heating altogether. This moves the seed survival curve to the right (Fig. 3.11C).

Few attempts have been made to quantify the variables presented in this model. In fact, there are generally too few data on pre-burn fuel conditions, fire behaviour and seed characteristics to permit accurate quantitative predictions (Sabiiti and Wein 1987). Floyd (1966) placed seeds at various depths in the soil profile in areas that were burned at three different fire intensities and then compared percentage germination of the exhumed seeds. This experiment showed the variation in depth of seed mortality in relation to fire intensity, as described in the above model. Moreover, it showed differences among species in the susceptibility to heating (Fig. 3.12).

These data introduce a further complication to the relationship between heating, depth, seed survival and germination, namely innate dormancy. Seeds exhumed from the deeper cachés failed to germinate even when tested in a germination cabinet at 26 °C, although they were still viable. This phenomenon of hard seededness is typical of many plant species in fire-prone ecosystems, and it is discussed in more detail below.

Mature seeds present on a plant at the time of fire may be protected simply by height in the canopy or by an insulating fruit. Surface fires in longleaf pine forests of south-eastern North America will usually have little impact on the canopy, where the current year's seed crop will be additionally protected in large cones. However, even in the so-called 'fire pines' (Warren and Fordham 1978; see Table 3.3) which suffer severe crown fires, seeds can obviously survive a fire and re-establish a post-fire population on the site.

Beaufait (1960) showed that single seeds of jack pine (*Pinus banksiana*) could tolerate a very brief burst of high temperature (i.e. 370 °C for 5–20

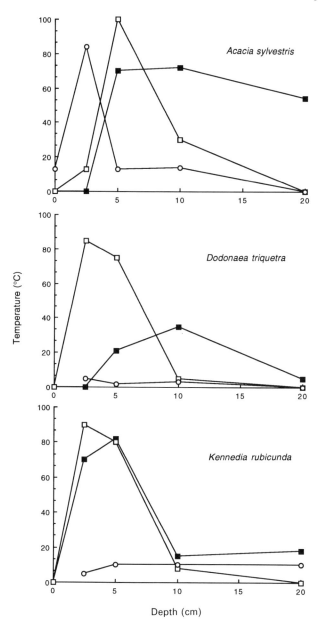

Figure 3.12 Relationship between germination of buried, dormant seeds and depth of burial at three different fire intensities: low (open circles), moderate (open squares), and high (solid squares), for *Acacia sylvestris*, *Dodonaea triquetra* and *Kennedia rubicanda* (modified from Floyd 1966).

s, or 430 °C for 5–10 s) without signficant loss of viability. On this basis, he inferred that individual seeds had evolved a substantial tolerance to high temperature, but this conclusion must remain dubious because there are very few data available for comparison. Perhaps all conifer seeds would tolerate this treatment. Linhart (1978) exposed seeds of three North American west-coast, closed-cone species to a range of temperatures from 85 °C to 125 °C for 5 min. Percentage germination remained high at 95 °C but declined by 105 °C for all species. The protective value of the cones was demonstrated by placing whole cones at 250 °C for the same time period (Fig. 3.13). Seeds in this treatment retained their viability almost to the level of unheated controls. This conclusion is supported in a study by Beaufait (1961), who found that, even in post-logging slash burns where fire intensity could exceed that generated in a wildfire, seeds on the remaining trees were sufficiently protected by the cones to produce a dense post-fire stand of seedlings.

Conifers are not the only plant taxon in which seeds are protected from the heat of a fire by being enclosed in woody fruits. The capsules of many eucalypts also protect the seeds from excessive heating in a fire. Striking examples also exist among woody perennial shrubs of the family Proteaceae in both Australia and southern Africa. The insulating function of the woody fruits of these species has not been measured directly, but it may be inferred from observations of seedlings appearing soon after fire, despite the fact that the fruits are borne at a height likely to suffer high temperatures for a substantial duration. Sonia and Heslehurst (1978) demonstrated that germination of seeds of three eastern Australian *Banksia* species is impaired at continuously-applied temperatures above about 25 °C, reaching zero germination at about 40 °C. Although these experimental conditions are nothing like those experienced in a fire, the temperature reached in the canopy of a burning *Banksia* shrub might be expected to reach 400 °C at 0.5 m and 200 °C at 3 m above ground (Bradstock and Myerscough 1981).

Several species of trees and shrubs retain annual seed production in fruits in the canopy, releasing them only after desiccation, following either death of the branch or fire, causes the fruits to open. For these species, the age of fruits stored in the canopy might be expected to influence their insulative function. Although vascular connections between branches and infructescences of *Banksia* plants are maintained for many years after flowering, older infructescences nevertheless appear drier than 1-year-old infructescences. This explains the observation that different degrees of scorching are sustained by the two age-classes of

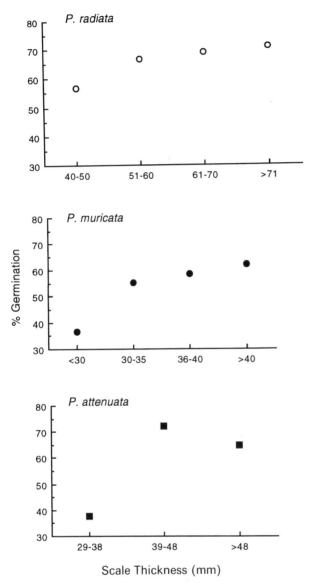

Figure 3.13 Comparison of percentage germination in relation to scale thickness of cones of three closed-cone pine species (data from Linhart 1978). Cones were exposed to 250 °C for 5 min in an oven, which is considered to be within the range of conditions experienced by cones in forest fire (see Beaufait 1960).

infructescences, even on the same branch at the same height (Fig. 3.14*A*). Whelan (unpublished data) found that a smaller proportion of seeds contained within these older infructescences was viable (Fig. 3.14*B*). Although drier fruits would transmit heat more rapidly to the enclosed seeds, reduced viability could also be explained simply by their greater age. Distinguishing between these possibilities requires further investigation.

It is clear from the indirect nature of the above examples that there are few experimental measurements of the capacity of cones and other woody fruits to protect enclosed seeds from heat damage in a wildfire. There are two principal problems with the information available. First, the heat treatments of seeds and/or cones are highly variable among studies, both in the temperature to which material is exposed and also in the duration of exposure (Table 3.1). It is therefore difficult to compare studies. We still have little idea of what constitutes a high degree of tolerance of heat for individual, exposed seeds. This needs further study. Second, laboratory heat treatments of fruits and seeds must be more closely related to the typical peak temperatures and durations experienced in a real fire. This requires a detailed knowledge of fire profiles in the natural vegetation, as discussed in Chapter 2.

Survival by seed mobility

There are plant communities in some regions in which a soil-stored seed bank appears to be completely absent (Johnson 1975, Thompson 1978), and there are plant species, in many communities, that do not sustain a dormant seed bank (Moore and Wein 1977, Ebersole 1989). Borrowing the reasoning from the vast literature on old–field, secondary succession or from studies on communities of intertidal animals, it is reasonable to identify a group of species that have their seeds escape fires by long-distance dispersal. *Epilobium angustifolium*, called 'fireweed' in many places, is a good example of this means of seed survival. As the common name suggests, this species flourishes after fire and many other sorts of disturbance (Myerscough 1980). Within northern hemisphere deciduous and coniferous forests, *E. angustifolium* is an important component of early successional communities and of early phases in regeneration cycles. Seeds retain viability for only a short while, up to 18 months (Granström 1987), so appearance of this species in profusion after fires must be by long-distance disperal from plants outside the area burned.

It is doubtful whether this life-history should be viewed as a highly evolved adaptation to fire, for any of the fire-prone ecosystems of the

A

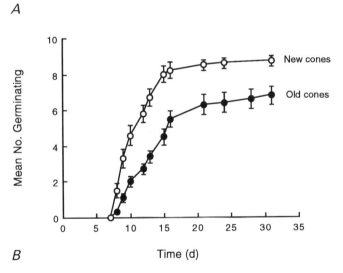

B   Time (d)

Figure 3.14   *A.* Infructescences of *Banksia paludosa* in eastern Australia exhibit different degrees of scorching depending on their age. Infructescences >1 yr old (on left) were thoroughly blackened by a wildfire which only scorched younger infructescences (on right) borne at the same height on the same plant.
B. Seed viability was related to age-class of infructescences (R.J. Whelan, unpublished data). Ten batches of 10 intact seeds were collected from 1-year-old (open symbols) and older (solid symbols) cones, germinated on wet filter paper, and germination was scored over a 1-month period.

Table 3.1. *Range of protocols used to determine seed viabilities and responses of seeds to high temperatures experienced in fires*

Plant species	Heat applied (°C)	Duration (min.)	Comments	Reference
Acacia sylvestris *Dodonaea triquetra* *Kennedia rubicunda*	40–100 at 5 °C intervals	10–800	Moist soil in oven	Floyd (1966)
Acacia spp. *Commersonia fraseri* *Phytolacca octandra*	40–100 at 10 °C intervals	10–400	Moist soil in oven	Floyd (1976)
Acacia suaveolens	40–170	1–360	Air-dried soil in oven	Auld (1986b)
Cassia spp. *Lespedesa stricta* *Desmodium tortuosum*	38–126	4	Dry heat and moist heat in oven	Martin & Cushwa (1966)
Cassia nictitans *Cassia aspera* *Lespedesa* spp.	80	0.25–16	Moist heat in oven	Cushwa et al. (1968)

Species	Temperature	Duration	Method	Reference
Cassia nictitans *Cassia aspera* *Lespedesa* spp.	45–110	?	Moist heat & dry heat in oven	Cushwa *et al.* (1968)
Stylosanthes spp.	25–50, 55, 65, 75	Diurnal cycles Continuous	Dry heat in oven	McKeon & Mott (1982)
Trifolium subterraneum	15–70 15–70	Diurnal cycles Continuous	Dry heat in oven	Quinlivan (1971)
Various Californian spp.	c. 100	Instant followed by cooling	'Hot water treatment'	Mirov (1936)
Pinus banksiana	370 370–538	1–3 0–0.3	Muffle furnace	Beaufait (1960)
Various west Australian forest spp.	105	2	Dry heat in oven	Bell *et al.* (1987)
Various spp. esp. legumes	100	Several	'Boiling water'	Rolston (1978) Bewey & Black (1982)
Various east Australian forest spp.	60–130 50–80	240 5	Dry heat in water	Beadle (1940)

world (Whelan 1986, Trabaud 1987a). This seems to be more of a generalist weed strategy, enabling colonization of areas denuded by any of a whole range of disturbances. However, the example does point to a question about the importance of effective local dispersal of seeds for some species in fire-prone environments, in particular, the group of woody perennial species labelled obligate seeders, which also have a canopy-stored seed bank and bradyspory (delayed dispersal). Established plants of these species are fire-sensitive, so the stored seeds are all-important in maintaining a population. A single fire may kill all the established plants in a population but will release the seeds from the seed bank. Assuming successful germination and adequate seedling establishment, the second fire is the critical one. If it occurs so soon after the previous fire that seedlings have not yet developed a seed bank themselves, it has the potential to cause local extinction. However, fires occurring at such a high frequency are likely to be patchy, because fuel loads are neither great nor uniform. Survival of seeds of obligate seeder species under these circumstances will depend on their having been dispersed to a patch that escapes the second fire.

The same importance of seed dispersal may apply to obligate seeders with a soil-stored seed bank faced with fires in quick succession. The genetic individuals that survive are either those derived from seeds that germinated in sites escaping the second fire, or seeds that had dispersed to a location in which they remained dormant and viable after the first fire.

Tolerating and exploiting post-fire conditions

The physical conditions after fire are markedly different from those that prevailed before, whatever plant community is burned. The physical effects of fire were described in Chapter 2. Of particular relevance are the alterations in soil temperatures, soil moisture levels, soil chemistry, insolation and nutrient availability. In addition, reduction in the sizes of animal populations may reduce the intensity of herbivory and seed predation. Individual plants may survive a fire and therefore experience these changed conditions, or their offspring may start life in the new conditions. How do they respond to the changed environment?

A wide variety of responses has been described (Table 3.2). These include: (i) increased productivity; (ii) increased flowering; (iii) improved post-fire seed dispersal; and (iv) improved seedling establishment. It must be emphasized that these responses are by no means universal across ecosystems nor among all species within an ecosystem.

Table 3.2. *Summary of reponses to burning exhibited by various plant species and possible explanations for these responses*

Observation	Possible explanations
Increased productivity	Increased nutrient availability Removal of suppressive dead leaves Increased average soil temperature Extended period of high temperatures Earlier start to growing season Removal of competing vegetation
Increased flowering	Increased nutrient availability Increase in numbers of shoots sprouting Removal of competing vegetation
Increased seed-dispersal distances	Removal of canopy from around fruits improves wind-flow Removal of ground vegetation and litter Greater foraging distances by seed-dispersal agents
Synchronous release of canopy-stored seeds	Heat treatment of sealed follicles or scales
Synchronous germination of soil-stored seeds	Heat treatment of impermeable seeds coats Charcoal residues break dormancy Alteration of surface light and/or temperature regime
Improved establishment of seedlings	Increased nutrient availability Decreased herbivore activity Satiation of populations of seed predators Removal of competing vegetation Degradation of allelopathic chemicals

Indeed, a single species may respond differently after different fires (e.g. p. 245 of Daubenmire 1968). An observed response may have both proximate and ultimate causes. For example, mass flowering of a sprouting plant species after fire may be a proximate response to injury, increased nutrient availability or post-fire soil temperatures. Alternatively, the mass flowering may be an evolved characteristic that maximizes fitness through increased pollination and/or satiation of seed predators (Gill 1981b, Kruger and Bigalke 1984), with the proximate factors used as cues. The proximate explanations are considered here, and a consideration of evolved characteristics is left until later.

Productivity

One effect of fire on plants that sprout after being burned appears to be increased vigour and growth rate. This phenomenon is well known in grasslands (Daubenmire 1968, Singh 1993), to the extent that there is a general belief that regular burning is required to maintain grass in a healthy condition (Kucera and Ehrenreich 1962). Observations of increased productivity also extend to forest trees, with reports of increased girth increments after some fires (Wallace 1966, Kimber 1978, Abbott and Loneragan 1983), and to forbs (Brewer and Platt in press), where greater increases in growth were noted after May (spring) fires than after fires in other seasons.

A temporary increase in productivity after fire in grasslands could have any of several reported causes (Table 3.2). The blackened soil surface may heat up earlier in the season, and earlier each day, than a comparable unburned site, and it may sustain higher temperatures through the day. For species in which productivity is limited by temperature, the total time suitable for plant growth may therefore be extended substantially. Adams and Anderson (1978) compared productivity of adjacent burned and unburned grassland in Oklahoma and related these differences to soil temperatures. Soil temperature was almost always higher in the burned site (Fig. 3.15A) and the peak of plant productivity occurred earlier in the year in the burned site and was nearly double that in the unburned site (Fig. 3.15B). Although this study did find a marked difference between burned and unburned sites, there was no replication of sites and the relationship of productivity to soil temperature is correlative. The weakness of this approach, which is very common in fire studies, makes it impossible to generalize about the effects of fire on productivity. It is therefore no surprise to find other studies that report no increase in productivity, and even a decrease, associated with fire (Wink and Wright 1973). Reasons for these differences need to be explored further, especially in relation to the timing of fire within a season, and to the interaction with rainfall and soil moisture. Properly replicated experimental studies are an important starting point.

The post-fire increase in the nutrients available to plants can stimulate plant growth in situations where nutrients set the limit to productivity. Daubenmire (1968) and Rundel (1982) both presented clear, compre-

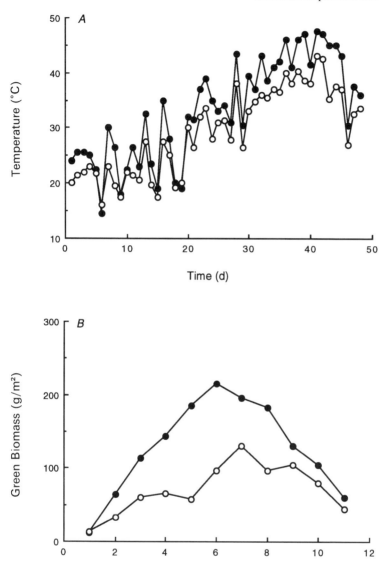

Figure 3.15 Relationship between soil temperature and productivity of grassland (from Adams and Anderson 1978). Both soil temperature (*A*) and plant productivity (*B*) were consistently higher in burned grassland (solid circles) than in unburned grassland (open circles) through spring and summer.

hensive reviews of nutrient availability in relation to fire in grasslands. In brief summary, foliar levels of nutrients are often found to be elevated after fire in many grassland ecosystems. Nevertheless, smaller amounts of nutrients are returned to the soil in ash after grassland fires, where biomass is low, than after forest and shrubland fires. The question to be addressed is whether measured increases in soil or foliar nutrients cause a post-fire increase in plant productivity. Experimental studies are few, and nutrient-addition or ash-addition experiments are needed to test the role of nutrient-return in productivity.

Lutz (1960) in North America and Wallace (1966) in Australia (see Fig. 3.16) both provided striking photographs of sections of tree trunks illustrating a spurt in growth, lasting for several years, following a major forest fire. This phenomenon is of obvious interest to foresters, and it is important to know whether the brief acceleration in growth is paid back by reductions some time later. For example, Henry (1961) found an increase in diameter increments over 5 years of annual burning in Queensland *Eucalyptus* forest, followed by such severe retardation in growth that, after 10 years, the burned trees had grown less than those in an unburned, control site. With the introduction of regular prescription burning as an integral part of forest management, it is also important to know whether acceleration in growth occurs after all types of fire, whatever the fire regime (e.g. frequency, season, intensity). Increased growth after even high-intensity fire is by no means universal (Hare 1961, Abbott and Loneragan 1983). Abbott and Loneragan (1983) found a significant long-term increase in diameter of Jarrah (*Eucalyptus margi-nata*) after a serious wildfire in Western Australia in only two of three sites and no increase or decrease was found in sites burned frequently in prescription fires.

Where increased productivity occurs, what is the cause? Several possibilities were suggested by Wallace (1966), including: (i) sudden increase in nutrient availability in ash; (ii) temporary removal of competition from understorey plants; (iii) thinning of tree stands caused by fire-induced mortality, particularly of suppressed trees; and (iv) the production of a dense crown of vigorously photosynthesizing leaves following epicormic sprouting. To this list, Kimber (1978) has added: (v) reduction in seed production for several years (Fig. 3.17). Once again, as for grasslands, studies providing evidence for each of these possibilities are mostly correlative and unreplicated. It is time to distinguish between the possibilities by manipulative experiments.

The observations of better post-fire growth, compared with 'control'

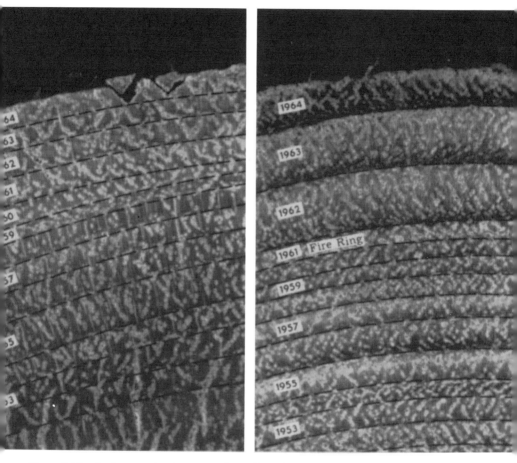

Figure 3.16 An increase in plant growth can be inferred from an examination of tree rings. Forest trees in Western Australia (Wallace 1966) show greater ring widths for several years after a wildfire in 1961 (right) compared with unburned trees (left).

sites burned less recently, may indicate a progressive inhibition of productivity in the absence of fire. Inhibition can be caused by an accumulation of litter and old stems. In a number of studies, experimental removal of litter from grassland sites has produced an increase in plant vigour comparable to that found after fire. Curtis and Partch (1950) found clipping of *Andropogon gerardi* swards produced increases in the density and height of flowering stems virtually equivalent to that occurring after burning. Moreover, burning followed by removal of ash and return of the litter layer obtained from an adjacent unburned

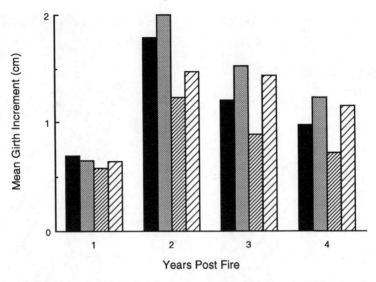

Figure 3.17   Annual growth increment for burned trees (solid bars) and unburned trees (hatched bars) which were either seeding (left-hand bar of each pair; $n = 37$ burned, $n = 88$ unburned) or not seeding (right-hand bar; $n = 63$ burned, $n = 77$ unburned). Data from Kimber (1978).

plot produced a flowering response equivalent to the unburned control sites.

Another interesting explanation for altered productivity is a change in soil chemistry over time. The process of *allelopathy* has been very appealing in attempts to explain post-disturbance changes in vegetation, especially in Californian chaparral communities (Muller *et al.* 1964, C.H. Muller 1965, W.H. Muller 1965, McPherson and Muller 1969). It is argued that toxins leached from leaves, produced by roots, or of microbial origin accumulate in the soil and inhibit germination of seeds or plant growth. A fire of sufficient intensity effectively "sterilizes' the soil and releases plants from the toxic element. In general, it is difficult to conduct experimental tests of an allelopathy hypothesis that are relevant to the field situation (see p. 372 of Harper 1977).

Flowering
The stimulation of flowering by fire has been more often reported than has improved growth, perhaps because a flush of flowering is readily apparent without precise measurement. A post-fire flush of flowering is more frequently reported for monocotyledons, particularly geophytic lilies and orchids (Gill 1981b, Kruger and Bigalke 1984, Le Maitre and

Brown 1992, Lamont and Runciman 1993), than for dicotyledons, especially shrubs and trees. Martin (1966, cited in Kruger and Bigalke 1984) listed the following five categories of flowering responses to fire in Africa, which could be applied equally well to other communities:

(i) immediate post-fire flowering but flowering very rare or completely absent if no fire;
(ii) immediate post-fire flowering with flowering less intense thereafter;
(iii) no flowering immediately after fire but flowering is intensified relative to unburnt sites after 3 or 4 years;
(iv) flowering depressed after fire, returning to levels comparable to unburned sites after some years;
(v) no apparent relationship of flowering to fire.

The causes of intense flowering may be closely related to increased productivity after fire – more vigorous plants may be able to support greater flowering (Daubenmire 1968, Rundel 1982). In addition, the damage caused by fire to above-ground parts of plants may stimulate the production of flower primordia. Thus, pruning is sometimes seen to have a stimulatory effect, producing an effect equivalent to that produced by fire. Gill and Ingwersen (1976) conducted a detailed study of the factors responsible for the spectacular post-fire flowering in the Australian grass-trees, *Xanthorrhoea* spp. (see Fig. 3.18). Burning doubled the proportion of plants that produced inflorescences, from one third to about two thirds. Clipping of leaves to mimic the pruning effect of fire produced the same response. Burning or clipping in spring stimulated flowering the following winter, whereas those untreated plants that did flower waited until the following spring and summer. Further exploration of stimulation of flowering, to examine the effects of season of burning or clipping (Gill 1981b), revealed a difference between these treatments. Clipping in winter did not appear to stimulate flowering in the following year: nor did burning in winter. However, winter burning did produce a flush of flowering in the second year that did not occur after clipping. Thus, there appears to be an interaction with the physiological status of the plant at the time of treatment.

In a similar sort of experimental study, but in South African fynbos, Le Maitre and Brown (1992) found that cutting of vegetation caused the same flowering response as burning in *Watsonia borbonica* (Iridaceae). Cutting of all vegetation except *Watsonia*, in treatment plots, caused a slight, non-significant reduction in the flowering response and cutting of

A

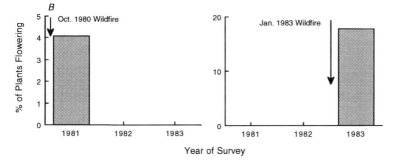

Figure 3.18   A. Intense flowering of *Xanthorrhoea resinosa* in the year after fire in south-eastern Australian sedgeland (photograph by P.M. Tap).
B. Percentage of *Xanthorrhoea resinosa* plants flowering in relation to time since last fire (from Jordan 1984). Data were collected from two sites, one burned in a wildfire in 1980 and surveyed for the three following years (left) and the other surveyed the year before and the year after a wildfire in 1983.

leaves from *Watsonia* alone did not stimulate flowering. The interpretation of this study is that flowering is stimulated not by fire itself, but by changes in factors such as soil temperatures, which are themselves affected by fire. Le Maitre and Brown also found that fire-stimulation of flowering in this *Watsonia* was much greater following summer/autumn fires than after winter/spring fires.

Seed dispersal
Very little work appears to have been done to examine seed dispersal after fire, within the burned area (see Whelan 1986). It seems likely that dispersal distances may well be enhanced after fire for some plant species. Surface wind and water flow are increased, and both these agencies have the potential to move seeds (see Chapter 2). For those bradysporous plant species (in which seeds are stored in the canopy, survive the fire and are released afterwards), seed dispersal distances could be greater due to the greater wind flow through the leafless canopy (Fig. 3.19). This suggestion requires further study, especially an investigation of the relative fates of seeds that disperse short and long distances from the parent.

Post-fire flush of seedlings
A post-fire flush of seedlings is frequently observed after fires, with seedlings occurring at much higher densities than in unburned vegetation. Increased seed release, increased germination and increased establishment may all contribute to this general observation.

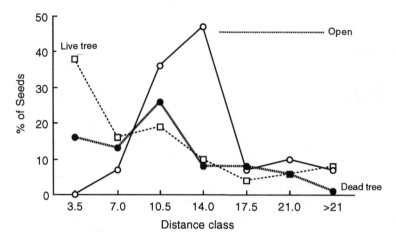

Figure 3.19 Patterns of seed dispersal from canopy-borne fruits of *Banksia ericifolia* in eastern Australia (R.J. Whelan, unpublished data). One hundred seeds were released from 3 m above the ground from each of three locations (wind speed 35 km h^{-1}): the top of a step ladder in the open (open circles); the top of a fire-killed tree which lacked foliage (black circles); the top of a live tree (open squares).

Delayed dispersal from a canopy-stored seed bank (bradyspory) is a mechanism that can produce a flush of seeds and therefore of seedlings. Some proportion of the seed bank may be released each year, because of spontaneous opening of a few fruits or death of a branch from breakage or insect attack. These could be classed as the 'soft' seeds in a cohort. The others are effectively dormant until the next fire causes widespread seed release. The bradysporous habit occurs in various taxa in several regions (Lamont *et al.* 1991b) – particularly the conifers in North America, Mexico and the Mediterranean (see Table 3.3); Cupressaceae, Casuarinaceae, Myrtaceae and Proteaceae in Australia; and the Proteaceae in southern Africa.

A semantic point is worth discussing here, because of the increasing use of the term serotiny to describe this phenomenon (see Le Maitre 1985 and Lamont 1991). Beaufait (1960) correctly identified the etymology of this word as *serus*, meaning 'late.' Botanists use the word *serotinous* to describe late-blossoming, and *serotinal* refers to the late-summer season of the year, especially used in descriptions of life-histories of freshwater organisms (Allaby 1985). This word therefore gives an inappropriate description of the phenomenon of seed release that is delayed, not just until late in the season but perhaps until the next fire many seasons hence.

Table 3.3. *Species of 'fire pines'. Cones in these species are sealed with resin, delaying seed release until after a fire has broken the resin bonds and allowed the cone scales to open. This table also includes reference to other studies on closed-cone pines and on other conifers, e.g. the Cupressaceae in California and Australia*

Region/species	Common name	Reference
Mediterranean region		
Pinus halapensis	Aleppo pine	Naveh (1974, 1975)
P. brutea		Naveh (1974)
P. pinaster	Maritime pine	
Tetraclinis sp.		Lamont et al. (1991b)
North America		
P. pungens	Table Mountain pine	Barden (1979)
P. rigida	Pitch pine	Givnish (1981)
P. clausa	Sand pine	Myers (1985)
P. serotina	Pond pine	
P. banksiana	Jack pine	Beaufait (1960, 1961)
P. contorta var. latifolia	Lodgepole pine	Perry & Lotan (1979)
P. attenuata	Knobcone pine	Linhart (1978), Vogl (1973)
P. contorta	Beach pine	Teich (1970)
P. muricata	Bishop's pine	Linhart (1978)
P. radiata	Monterey pine	Linhart (1978)
Cupressus forbesii	Tecate cypress	Zedler (1977)
Other Cupressus spp.		Lamont et al. (1991b)
Mexico		
P. greggii	Gregg's pine	
P. oocarpa	No common name	
P. patula	Spreading-leaved pine	
P. pringlei	Pringle's pine	
Australia		
Callitris spp.	Cypress pine	Lamont et al. (1991b)
Actinostrobis sp.		Lamont et al. (1991b)

Source: Based on Warren and Fordham (1978) and Lamont et al. (1991b).

Bradyspory is the more appropriate term (Specht 1979, Kruger and Bigalke 1984, Pate and Beard 1984) and it is used throughout this book, notwithstanding the fact that the usage of 'serotiny' predates 'bradyspory.' Lamont et al. (1991b) distinguished between bradyspory (serotiny), which is the phenomenon of seed retention, and 'pyriscence' which means that seed release is stimulated by fire.

Fire can be a double-edged sword for seeds that have been dispersed and are incorporated in the soil/humus. On one hand, seeds on the surface may be consumed by the fire or may at least be killed by the burst of heat. In most plant communities, however, there is an enormous dormant seed bank present in the soil and it is clear that fire can stimulate germination from this seed bank. There appear to be several mechanisms by which stimulation of germination can operate, including ageing of the seed, passage through the gut of a vertebrate, physical scarification or unsuccessful predation attempts. Fire provides an altered physical environment, with increased light intensity and altered quality at the surface and heated soil stimulating germination. Another stimulus with a more obscure mechanism, operating for seeds with innate dormancy enforced by factors other than an impermeable seed coat, may be chemicals leached from charcoal (Keeley et al. 1985, Keeley 1986b). However, heat generated by the fire is perhaps the most common stimulus for a range of species. Seeds of many species in several families (e.g. Leguminosae, Cannaceae, Malvaceae, Convolvulaceae – Bewley and Black 1982; Leguminosae, Sterculiaceae and Anacardiaceae – Mirov 1936) have a form of dormancy enforced by an impermeable seed coat. The hard seed coat may itself confer some tolerance of heat (Beadle 1940), but it must be broached somehow, if germination is to occur.

Successful germination of soil-stored, dormant seed is therefore a fine balance between suffering mortality caused by excessive heat, as discussed above, and receiving sufficient heat to break dormancy. Floyd (1966) and Auld (1986b) have examined this problem in some detail, varying the temperature applied to seeds and the duration of exposure, for a number of Australian hard-seeded species. Their results (Fig. 3.20) illustrate the increased complexity that seed dormancy adds to the model presented in Fig. 3.11. The proportion of seeds germinating successfully will be a function not only of depth but also of the degree to which dormancy is broken by heat.

For some plant species, a flush of germination may follow in the year following fire, when the established plants have sprouted, flowered and set seed (Kruger and Bigalke 1984, Whelan 1985). In such cases, dense germination, compared with that at unburned sites, results directly from enhanced flowering following fire. In a single plant community, the post-fire flush of germination may be made up both of species with seed release and germination triggered by fire (predominantly in the first year after fire) and also of species that sprout, flower and release non-dormant seeds (predominantly after the first post-fire year).

Increased seedling establishment and growth may be related to the

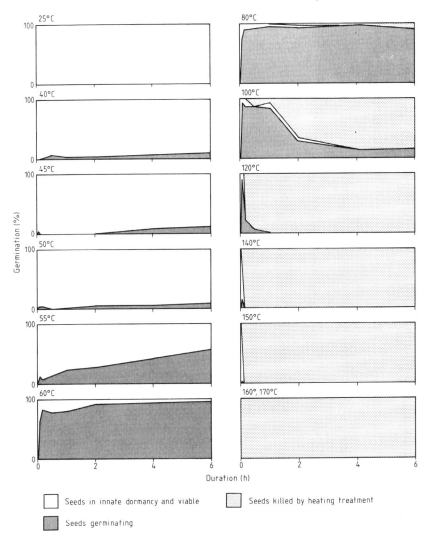

Figure 3.20 Fates of *Acacia suaveolens* seeds exposed to a range of temperatures and durations (reproduced from Auld 1986b with permission of the author and Blackwell Scientific Publications). At low temperature (25 °C), germination was 0 and all seeds remained in innate dormancy (no shading). At higher temperatures (up to 80 °C), germination increased (dark stippling) with increasing temperature and duration, with the remaining seeds remaining dormant. At higher temperatures still, and long durations of high temperature, most seeds are killed (light stippling). A 40-min. heat treatment of 40 °C and 120 °C would therefore produce similar germination rates (c. 10%), but at 40 °C the remaining seeds would remain dormant and viable whereas at 120 °C the remainder would have been killed.

wide range of beneficial factors discussed above under 'Productivity', especially increased availability of nutrients and reduced competition. Herbivory may be temporarily reduced. To balance these, however, post-fire physical conditions may also be detrimental to establishment and growth of seedlings, with severe high temperatures and reduced water availability causing death. Thus, the occurrence of mass germination after fire may depend on the coincidence of the timing of release of seeds from dormancy and favourable post-fire climate. Bond *et al.* (1984) and Bond (1984) reported that numbers of seedlings of several species of South African Proteaceae establishing were much greater after autumn than winter or spring fires. Similar results were obtained for species of Proteaceae in another Mediterranean-climate region – Western Australia (Cowling and Lamont 1984, 1987). There are two main explanations for low seedling establishment apparent from these studies (see Midgley 1989, Bradstock and Bedward 1992). First, those seedlings that appear after a spring fire suffer desiccation over the following hot, dry summer. Second, dispersed seeds that remain in enforced dormancy over summer, requiring adequate moisture and cool temperatures to stimulate germination, are exposed to a high risk of mortality from seed predators or lethal temperatures (Fig. 3.21).

Whelan and Tait (1995) suggested that this seasonal pattern is by no means general, and is perhaps confined to the strongly seasonal Mediterranean climate regions such as South Africa and Western Australia. In experimental studies in eastern Australia, near Sydney, in which marked seeds were placed in replicated plots burned in different seasons, they found that germination occurred soon after one autumn fire but was delayed for over a year following another autumn fire.

Post-fire inhibition of germination

Poor germination and seedling survival in some areas has been attributed to the physical and chemical effects of heating of soils. Heating of moist soil can drive some chemicals down the soil profile, producing a water-repellent layer that enhances runoff and inhibits rainfall soaking in to the soil (DeBano *et al.* 1967, DeBano 1971, Scott and van Wyk 1992).

Soil sterilization has also been noted as an effect of fire, occurring especially in locations where fire intensity is high and sustained for a long time, such as under burning logs, in piles of timber on cleared land and under post-logging slash. Soils are mineralized under these conditions, and after fire, the soil surface can be colonized by lichens and mosses but little else for decades (pers. obs.).

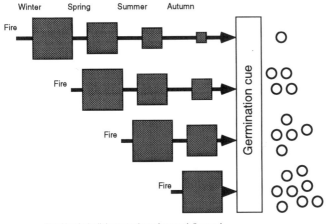

Figure 3.21 The timing of fire could influence successful establishment of seedlings for various reasons. This schematic diagram (modified from Bond 1984) illustrates the possible impact of post-dispersal seed predators. The shaded boxes on each arrow represent the change in seed bank after a fire in a particular season. The size of the squares represents the magnitude of the bank of *viable* seeds remaining. Late autumn represents the cue for germination, so the earlier a fire, the greater the depletion of seeds by late autumn.

In soils of poor nutrient status, mycorrhizal fungi can be critical root symbionts, enhancing water and nutrient uptake of seedlings and established plants (Harley and Smith 1983), and it might be expected that the heat of fire could reduce mycorrhizal fungi populations (Klopatek *et al.* 1988, Wicklow-Howard 1989). Dhillion *et al.* (1988), working in prairies, found that both colonisation of roots by vesicular arbuscular mycorrhizal (VAM) fungi and sporulation of the fungi were reduced after fire. This was related to lower above-ground plant production and lower inorganic nutrient content. In contrast, Bellgard *et al.* (1994), working in a sandstone community near Wollongong, found no effects of a wildfire on VAM colonization of the roots of a bioassay plant, *Acacia linifolia*.

Categories of response to fire

The wide variety of plant responses to fire discussed above can be summarized in a number of alternative classifications, which Gill and Bradstock (1992) have referred to as: 'mechanistic' (referring to the mechanisms by which individual plants may tolerate a fire – e.g. thick bark), 'consequential' (referring to the consequence of the fire in terms of alteration to a plant population); and 'strategic' (referring to 'behaviours' of plants that might be viewed as the evolutionary consequences of long exposure to fires). A plethora of terms has sprung up to describe various responses in each of these classifications, with little uniformity of common use. The following three categories are probably adequate: *fire ephemerals, obligate seeders*, and *sprouters*, with several subcategories of sprouters defined if necessary (Bell *et al.* 1984). The following discussion relates the variety of classifications in the literature to these preferred categories (Table 3.4). An interesting feature of different terms used to describe the same biological response of a plant species to fire is that some are based on the patterns of re-establishment after fire, in order to explain changing community structure, and others are based on patterns of survival and mortality of individual species.

A clear major division among the species listed in Table 3.4 relates to the ability of established plants to tolerate fire either by protection of aerial buds or by sprouting from protected buds in the stems or roots. This forms the major division, defined in 1974 by Biswell as *non-sprouters*, which die in a fire, and *sprouters*, which survive. There is an important problem with this apparently clear classification. A species may respond differently to two fires at the same site, for any of several reasons, including different seasons of burning, different intensities of fire, or because the fire under observation suffered a previous fire too recently. Gill (1981b) attempted to take this into consideration by creating a functional definition of capacity to sprout in relation to plant age and to fire intensity as perceived by the plant. Thus, a species in which mature plants can recover after 100% leaf scorch (as distinct from complete combustion of leaves) would be classified as a sprouter. The need for a functional definition such as this occurs because survival of individuals of what could reasonably be defined as a sprouter species is not usually 100%. Keeley and Zedler (1978), for example, reported that mortality of shrubs after a chaparral fire in California was 100% for the non-sprouter species but variable in the sprouting species, ranging from none in some species to high in *Adenostoma*. The national register for the fire responses of plant species in Australia, being established by Gill and Bradstock (1992), is an attempt to clarify fire responses and will permit the identification of intra-specific variation in these across plant communities.

If established plants of a species die in a fire, how does the species

Table 3.4. *Comparison of various schemes for classifying responses of plant species to fire*

Biswell (1974)	Keeley (1981)	Kruger (1977)	Naveh (1975)	Rowe (1983)	Gill (1981b)	Noble (1981)
Non-sprouters	Obligate seeders	Reliant on seed production	Obligatory seed regenerators	Disseminule-based: Invaders Avoiders	Non-sprouters: No seed storage in burnt area	Propagule-based: Dispersed propagules
				Evaders	Seed storage on plant Seed storage in soil	Long-lived propagules Long-lived but entire germination in fire Short-lived propagules
Sprouters	Sprouters	Resistant to scorching		Vegetative-based: Resisters	Sprouters:	Vegatative-based:
						Unaffected by fire Adults unaffected juveniles killed
		Vegetative regeneration	Obligatory resprouters	Endurers	Subterranean regeneration: basal stems, root suckers & rhizomes	Resprout via juvenile stage
			Facultative root resprouters	Combination		

regenerate in a burned site? Incorporating the answer to this question into a classification has produced a wide range of terms, including *obligate seeder, obligatory seed regenerators*, species *reliant on seed production*, and *disseminule-based* and *propagule-based* regeneration. Some of these are further subdivided to define characteristics of the seed source, such as where it is stored and whether it is long-lived or short-lived. Similarly, species in the broad *sprouter* category have been further subdivided on various grounds, including where sprouting emanates (above or below ground), and whether juvenile plants are affected even if adults survive.

The recognition that a large group of sprouter species, particularly woody perennial shrubs in Australia and southern Africa, have their germination tied to fire in some way has led to an additional category (*combination* or *facultative root resprouter*), shown at the bottom of Table 3.4.

What is the value of these sorts.of classification? They are used in two main ways: (i) to compare the structure of plant communities in different regions (e.g. Lamont *et al.* 1985), and (ii) to generate predictive, population-dynamics models of plant community change in response to an altered fire regime (e.g. Noble and Slatyer 1980). The evolution of different modes of response to fire is considered later in this chapter.

Tolerance of animals to fire

Direct exposure to fire

Animal cells are damaged by high temperature. At the extreme, incineration can be the fate of an animal overtaken by the fire front. Anecdotal reports of charred carcases found after wildfires are manifold (Chew *et al.* 1958, Ahlgren and Ahlgren 1960, Bendell 1974, Newsome *et al.* 1975). However, little is known of about the physiological effects of the short burst of heat in a fire on animals in the field. It is not an easy area for study. Some factors that have been suggested as contributing to heat death in animals are:

(i) Denaturation of proteins;
(ii) Thermal inactivation of enzymes faster than they can be re-formed;
(iii) Inadequate oxygen supply;
(iv) Differential temperature effects on interdependent metabolic reactions;
(v) Temperature effects on membrane structure.

Most of these explanations, discussed in more detail by Schmidt-Nielsen

(1979), pertain to the situation in which the animal's body tissues heat up in response to high ambient temperature. It is apparent that, even for animals exposed to sustained high temperatures, the ultimate cause of heat death is poorly understood, and seems likely to comprise several of the above explanations. Nevertheless, the highest body temperature at which animal life is considered possible is about 50 °C.

The situation for an animal faced with a bushfire is considerably different from that of an animal living in an environment of sustained high ambient temperatures. A very much higher temperature would be experienced but only for a short time. Although few experiments have directly examined the effects of such a treatment, there are several principles we might apply to make reasonable predictions. As for plants, we can ask whether critical tissues are protected from high temperatures.

Of course, animals are basically different from plants in that few terrestrial species are 'modular'. A serious injury to some part of an animal usually means the death of the whole. A plant, in contrast, may suffer extreme injury to some branches, or even to all above-ground parts, and still be capable of recovery. The importance of this seemingly trivial observation is that a fire can cause death of an animal by direct injury, without necessarily heating internal body tissues to a lethal temperature. The kangaroos and wallabies seen hopping around 'with sore feet' after an intense fire in south-eastern Australian heath (Newsome et al. 1975) may have died subsequently, despite avoiding direct thermal stress and surviving the passage of the fire front.

Resistance to heating

The ectoderm provides the first barrier to radiant heat reaching underlying tissues, and three factors will determine the protection of critical body tissues from lethal temperatures: (i) the insulating capacity of the ectoderm: fur is a better insulator than skin; (ii) the size of the organism; and, interacting with these two factors, (iii) the duration of exposure to heat.

Studies on the insulation value of fur have mostly been confined to tolerance of cold temperatures (Schmidt-Nielsen 1979). However, the thermodynamic principles are the same whether heat is moving through fur into or out of the animal. Thermal conductivity (k) is measured in units of heat transfer, per unit of time, per unit of surface area, per unit of temperature differential (e.g. cal s^{-1} cm^{-2} °C^{-1}). Measurements of thermal conductivity on a variety of materials show that animal fur has lower conductivity (i.e. is a better insulator) than tissue ($k = 9.1 \times 10^{-5}$

versus $k = 1.1 \times 10^{-3}$) (Table 8.3 of Schmidt-Nielsen 1979). Therefore, for a particular temperature differential and animals of equivalent surface area exposed to heat, a mammal will heat up more slowly than a reptile, for example, which lacks fur. The insulating capacity of fur is generally related to its thickness and to the amount of air trapped within it. Furthermore, there is a correlation between mammal size and fur thickness. Thus, small mammals will have poorer insulation solely because they typically have shorter fur than larger mammals.

The dice are loaded against small animals. Small body size is a disadvantage because of what might be called 'thermal inertia'. Schmidt-Nielsen (1979) provided an elegant analogy:

If we put a block of ice ouside in the hot sun, and a small piece next to it, the small piece will melt away long before the large block. If a big rock and a small pebble are placed on the ground in the sun, the pebble will be hot long before the big rock. The reason is that a small object has a much larger surface area relative to its volume.

Perhaps a fire-front passes so rapidly that insulation such as fur, and the thermal inertia of a reasonable body size, may be protection enough: this needs further study. There have been many observations of large mammals such as ungulates and macropod marsupials doubling back through the fire front. Large size and fur together may make this brief encounter with very high radiant heat possible. However, animals that avoid the extreme temperature by taking refuge in burrows or tree hollows may experience elevated temperatures for extended periods. Lawrence (1966) measured temperatures in various locations during a fire in Californian chaparral. Although peak surface temperatures exceeded 300 °C, the peak temperature at 15 cm down a straight burrow was only 72 °C. This is an impressive difference, but the temperature at this depth did remain above 50 °C for about 20 min. Under such conditions, large body size with its associated themal inertia and perhaps thicker fur will not provide protection, because even a large animal has time to heat up to a lethal temperature. The ability of an organism to hold down its temperature by thermoregulatory mechanisms will be important in survival.

Thermoregulation
In general, thermoregulation may be achieved either through behaviour, by seeking out suitable microsites and by adopting suitable postures, or through physiological mechanisms, principally evaporative cooling (Schmidt-Nielsen 1979). Sheltering immobile in a refuge, with

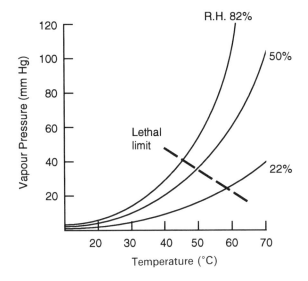

Figure 3.22 Relationship between vapour pressure and ambient temperature at different humidities (82%, 50%, 22%) (from Lawrence 1966). Because of this relationship, low humidity permits animals to tolerate much higher temperatures than if humidity is high (see the intersection of lethal limit line and humidity curve).

the ambient temperature rising and remaining high, constrains the behavioural mechanisms of thermoregulation that can operate when radiant heat comes from a particular direction. Evaporative cooling seems the only possible means of survival where ambient temperatures exceed the lethal temperature for an extended time.

The effectiveness of thermoregulation by evaporative cooling is dependent upon relative humidity. Lawrence (1966) pointed out that vapour pressure is an important expression of the capacity of an organism for evaporative cooling, but the relationships between temperature, humidity and survival in the field are not well studied. Howard *et al.* (1959) and Lawrence (1966) conducted experiments that would attract the scrutiny of animal ethics committees today! Caged animals were placed in various locations in grassland/shrubland and chaparral respectively, prior to ignition of experimental fires. These studies produced the model shown in Fig. 3.22, for small mammals in California chaparral. Vapour pressure rises much more rapidly with increasing temperature at high humidity than at low humidity. The limited amount of data collected in the field 'survival' experiment suggests that the lethal limit of temperature would be about 60 °C at 22% relative humidity but only about 45 °C at 82% relative humidity.

If animals can survive elevated temperatures within the refuge by evaporative cooling, their continued survival depends upon the ability to repay the water debt. Thus, they must have access to some source of water after the fire. Free water is likely to be scarce in arid and semi-arid environments when fire occurs in a dry season. The above discussion relates mostly to mammals. Do ectotherms have the same problems? Substantial differences might be expected, because the lack of hair as an insulating layer would mean that these animals would heat up more rapidly. However, having a lower metabolic rate than equivalent sized mammals may reduce the urgency, in ectotherms, of paying back a water-debt imposed by thermoregulation during the passage of the fire. Few field experiments appear to have been conducted in this area (Friend 1993).

Smoke and anoxia

Even if the air temperature surrounding the animal does not reach lethal levels, smoke inhalation can be of great importance (Bendell 1974). Clear examples of this are found in many reports of injuries received by firefighters of both wildfires and city fires. Medical treatment for smoke inhalation is far more common than treatment for burns. Zikria *et al.* (1972) reported that 76% of deaths in fires in New York City were due to some type of respiratory failure (e.g. carbon monoxide poisoning, lack of oxygen, toxic chemicals in smoke). Christensen (1980) reported that the death, in an experimental fire, of a single woylie (*Bettongia penicillata* – a medium-sized marsupial) was due to suffocation in the hollow log in which the animal had sought refuge. Lawrence (1966) suggested that rodents (*Peromyscus*) in burrows survived the passage of fire if the burrow had two openings, allowing ventilation, whereas animals in burrows with only a single entrance apparently suffocated.

Suffocation can also be caused by low oxygen availability. The pyrolysis reaction consumes oxygen, and could reduce ambient oxygen levels for long enough to cause death by asphyxiation. Once again, differences between endotherms and ectotherms would be expected, because differences in metabolic rate may allow ectotherms to tolerate anoxic conditions for longer periods.

The importance of time

The above discussion clearly indicates that 'time is of the essence' for animals caught in a fire. Transfer of radiant heat from the fire into an animal's body is a time-dependent phenomenon. Thus, as long as the ectoderm of an animal is dead tissue, or can be replaced after injury, underlying tissues will not reach a lethal temperature if the temperature differential between the body and the environment lasts for only a short time. The greater the temperature differential (i.e. the more intense the fire front), the faster the rate of heating.

Avoidance of direct heat

Behaviour in the face of fire

Peak temperatures reached on the soil surface in many types of fire are usually well in excess of the lethal temperatures recorded for animals (about 50 °C). Nevertheless, many animals escape incineration, death through heat stress and injury, and asphyxiation. It therefore appears that behavioural characteristics that permit avoidance of high temperature for sufficient time are widespread, and frequently explain fire tolerance of animals.

Our anthropocentric view of the world leads to the common expectation that animals will display widespread panic in the face of fire. This view is wonderfully illustrated in the much thumbed postcard shown in Fig. 3.23, found in a North Wales seaside resort in the UK! However, observations made in the vicinity of an advancing fire, in many different ecosystems in several continents, present a different picture (Komarek 1969, Vogl 1973, Bendell 1974, Main 1981). Large, mobile animals, in particular, seem capable of moving quickly to unburned parts of the habitat or back throught a break in the flames to relatively safe, burnt ground. Christensen (1980) studied the behaviour of a small kangaroo-like marsupial, the woylie (*Bettongia penicillata*), during a fire of moderate to high intensity in south-western Australian eucalypt forest. The animals under observation all remained within their home ranges, which had been determined prior to the fire, staying within the vicinity of the nest area (Fig. 3.24). Nine of the 19 woylies that were radio-tracked successfully doubled back through the flames to remain within their home ranges, as illustrated by the animal shown in Fig. 3.24. Six animals found an unburned refuge within the home range and four animals sought refuge in hollow logs (Recher and Christensen 1981).

Not all animals show such an attachment to a home site and may simply move before a fire front until they end up in a patch of vegetation that remains unburned. There are two lines of evidence for this. The most frequent is the observation that animals found alive immediately

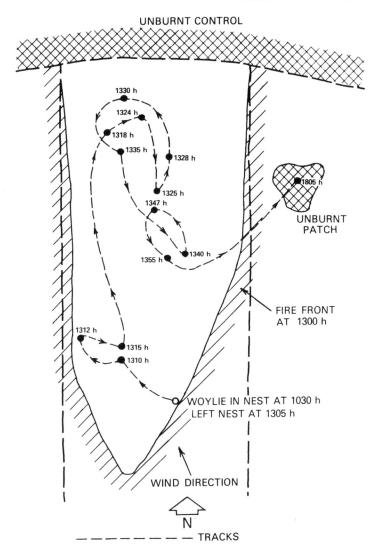

UNBURNT CONTROL

1330 h
1324 h
1318 h
1335 h
1328 h
1325 h
1347 h
1340 h
1355 h

1805 h

UNBURNT PATCH

FIRE FRONT AT 1300 h

1312 h
1315 h
1310 h

WOYLIE IN NEST AT 1030 h
LEFT NEST AT 1305 h

WIND DIRECTION

N

— — — — — — — — — TRACKS

Figure 3.23 (left) The common perception of the likely impact of forest fires on animals: a complete catastrophe causing panic and flight (from a postcard found in a North Wales seaside resort, UK).

Figure 3.24 (above) Movements of a woylie (*Bettongia penicillata*; macropod marsupial) as a fire front approached during an experimental, high-intensity fire in a south-west Australian eucalypt forest (from Christensen 1980). The animal remained within its home range, doubling back through the fire front and into a patch of vegetation left unburnt in the fire. Cross hatching represents unburned vegetation; hatching represents the burned side of the fire front.

after wildfires are in locations that may be seen as refuges. Studies of small-mammal populations after fire, for example, report survival of animals in unburnt patches of vegetation such as wetter streamside areas (Newsome et al. 1975) or rocky outcrops (Lawrence 1966). There are also reports of many organisms seeking out, and surviving in, burrows and crevices; for example, small reptiles and frogs (Main 1981). Observations such as these may indeed imply that the animals were able to seek out unburned refuges, but the same observations could alternatively be explained by the possibility that all animals were killed except those within the patches of vegetation that happened to escape the fire.

Surprisingly few studies have attempted to separate these two possible interpretations, but there is some support for the former. Tevis (1956) tagged rodents (*Peromyscus* spp.) before a fire in post-logging slash in Douglas Fir forest in North America. Tagged animals were recovered only in unburned patches within the burned area and in surrounding unburned forest. Gandar (1982) sampled grasshoppers in African savanna before and after a fire in both burned and unburned patches of vegetation. Total biomass of grasshoppers in the unburned patches increased, immediately after the fire, to three times pre-fire biomass while biomass in burned areas declined dramatically. These results indicate survival of many individuals and a movement from burned savanna to the unburned patches.

Unburned patches of habitat within the burn area are not the only potential refuges for animals in a fire. J.C. Taylor (pers. commun.) reported mass movement, in Western Australia *Eucalyptus* forest of litter-dwelling arthropods before a fire front; some of these animals were recorded burrowing into the dense crowns of *Xanthorrhoea* plants (Main 1981) (see Fig. 3.7). Whelan et al. (1980) compared the invertebrate inhabitants of plants with dense crowns immediately after a wildfire with similar plants in an unburned site. The results suggested strongly that the animals were seeking out the crowns of these plants as potential refuges (Table 3.5). Gandar (1982) also reported this behaviour for grasshopper nymphs, which survived fire in African savanna by burrowing into grass tussocks, where they were likely to be somewhat protected from the heat of a fire. For example, D. Gillon (1972) measured temperatures within burning grass tufts in East African savanna, and reported that maxima were as low as 50 °C. Other measurements of peak temperatures in potential refuge sites within a burn would be valuable.

It is important to know whether refuges hold a concentration of

Table 3.5. *Collections of invertebrates from the crowns of* Macrozamia riedlei *and* Xanthorrhoea preissii *plants in adjacent burned and unburned areas of south-west Australian eucalypt woodland*

Taxon	Macrozamia		Xanthorrhoea	
	burned	unburned	burned	unburned
Hymenoptera				
Rhyditiponera sp.	x	—	—	—
Iridomyrmex sp.	x	—	—	—
Iridomyrmex sp. 2	—	—	—	x
Polyrachis sp.	—	—	x	—
Coleoptera				
Nitidulidae sp.	x	—	—	—
Adelium sp.	—	—	x	—
Lycidae sp.	—	—	x	x
Paropsis sp.	—	—	—	x
Blattoidea				
Blattidae sp.	x	—	x	—
Blattidae sp. 2	x	—	x	x
Hemiptera				
Reduviidae sp.	—	—	x	—
Aranaea				
Lycosa sp.	x	—	x	—
Salticidae sp.	—	—	x	—
Thomisidae sp.	—	—	x	—
Theridiidae sp.	—	—	x	—
Myriopoda				
Cryptops sp.	—	—	x	—
Total number of taxa	6	0	11	4

Notes: x = species recorded in the sample.
Source: From Whelan *et al.* (1980).

animals after fire, or simply just the remnants of populations at their original densities. One reason is that high densities of herbivores within unburnt patches of vegetation will surely lead to severe grazing pressure (see Chapter 7). Moreover, there will be a reservoir of herbivores able to start feeding within the regenerating burnt area as soon as sprouts and/or seedlings appear. If refuges hold solely their share of the pre-burn populations, grazing pressures will not be elevated within the refuge and grazing in the surrounding burnt vegetation will be deferred until breeding provides colonists.

As a postscript to the above discussion about the capacity of many animals to escape the direct effects of fire, Chandler *et al.* (1983, Vol. 1 p. 208) sounded a word of warning about making general conclusions. They argued that the widely held belief that wildfire is not a serious cause of mortality in vertebrates may be misplaced, because it has been put forward by researchers whose experience comes from studying low-intensity, prescribed fires and small-area wildfires. This possibility seems quite likely, and it is therefore important to pursue it. To do so will require challenging studies of individual, tagged vertebrates before and after large-area, high-intensity experimental fires.

Life-history as a defence

Physiologically, resting stages of an organism's life cycle are likely to be especially tolerant of heat (p. 210 of Schmidt-Nielsen 1979), especially where they are also dehydrated. Thus, in climates where the likely wildfire season corresponds to a hot, dry season, many animal species may survive because they are in an aestivating stage, and thereby more tolerant of high temperature and perhaps also buried – a double protection. Individuals may also survive a fire that occurs during another tolerant part of the life cycle. Y. Gillon (1972) and Gandar (1982), for example, suggested that grasshoppers species survived a September (dry season) fire in east African savanna because most individuals were adults at this time, and are therefore able to flee. Kruger and Bigalke (1984) described other examples of South African insect species with specialized life cycles that minimize susceptibility to fire. In addition, they provided one of the few published cases, for a vertebrate, of avoidance of fire by life-history. *Psammobates geometricus* (the geometric tortoise) lays eggs in spring, buried about 10 cm below the soil surface. Incubation lasts throughout the fire season, until April–May, when the likelihood of fire becomes low again. Winter and spring provide the best forage and support rapid growth rates of the young tortoises, which are well established by the first fire season. These animals avoid close, mature vegetation, which has remained unburned for many years, favouring the more productive, open, recently burned vegetation. These two characteristics – the timing of egg laying and hatching and the habitat selection – suggest two levels of fire avoidance: eggs are protected from most fires likely to burn a site; and the animals spend most time in vegetation that is least likely to carry a fire.

Effects of changed post-fire conditions

Mobility of animals and variability in fire behaviour combine to allow many animals to survive the passage of the fire front. These animals are now faced with a much altered environment. The hostility of this post-fire environment probably explains the common observation that the animal species present before a fire may be recorded immediately afterwards but soon disappear. Perhaps of greatest significance is the removal of plant biomass that previously provided food and cover.

The fundamental importance of the vegetation as animal habitat is well recognized by zoologists. The vegetation either provides directly, or moderates, virtually all of the habitat requirements of animals – physical conditions of temperature and wind, cover from predators, nest sites, and food. Identifying the habitat requirements of a particular animal species, and knowing the way in which the vegetation will change after fire, should permit a prediction of the response of the animal to the fire.

Kruger and Bigalke (1984) provided a clear example of the effect of fire on an animal species via alteration of the provision of food and nest sites by the vegetation. The sugarbird (*Promerops cafer*) in African fynbos relies on the elevated inflorescences of shrubs in the family Proteaceae for the supply of nectar and insects. Most of these plant species are obligate seeders; that is, the adult plant dies in the fire and a population re-establishes from seed. The new individuals do not reach reproductive age for 4–8 years, leaving the birds without a food resource for this time period. Added to this, the birds require a dense, sheltering canopy for nest sites, and this does not re-form for about 8 years after fire.

The importance of food availability is reflected particularly in many studies of the responses of small mammals to fire. For example, Fox and Fox (1987) reviewed studies demonstrating the importance of re-establishment of the litter layer, and its associated invertebrate fauna, for recolonization of burned *Eucalyptus* woodland by native rodents in eastern Australia. The study of woylies by Christensen (1980), described above, showed that the individuals that survived the experimental fire and remained within their home ranges altered their diets substantially. Hypogeous fungi became a predominant item in the diet, where plant material had formerly been most abundant. Plasticity in diet is not unique to Australia nor to marsupials. A group of macaques in Borneo responded to extensive wildfires in East Kalimantan in 1983 with marked changes in diet from fruits, seeds and flowers before the fire to desiccated fruits, herbs, and insects afterward. These changes were accompanied by changes in behaviour, with more travel on the ground rather than through tree canopies, and less group adhesion (Berenstain 1986).

There are many animals, especially burrowing forms, that appear to be able to survive the passage of a fire, and to find sufficient food afterwards. This may not be enough to ensure survival, because predation may be much more severe after fire, when vegetation cover has been removed and access to predators is facilitated. Christensen (1981) discovered that predation accounted for the loss of several woylies, after most had survived the fire and were able to feed on an alternative food source.

The physical and chemical changes wrought by fire (Chapter 2), especially redistribution of plant nutrients and increases in their availability, might be expected to improve the quality of regrowth as forage. This was well recognised by early farmers in several continents. For example, early settlement in Australia included cattle grazing and this was almost inevitably associated with deliberate burning – 'To support small herds of livestock, the grassy woodlands surrounding the fields were burned' and 'Cattle refused to eat except on the burned sites' (Ch. 13 of Pyne 1991). Fire therefore appears to improve the quality and quantity of food available to herbivores. Some predators might be expected to benefit too, as the more open post-fire vegetation is easier to move through and prey may be more readily detected. Some animals might benefit from fire because of a reduction in parasitism. Bendell (1974) described studies of parasite loads in a population of blue grouse 5 yr and 12 yr after a wildfire. The results are striking – 17% of animals were free of blood parasites at 5 yr but all were infected at 12 yr. Frequencies of infection with particular blood, gut and external parasites at 12 yr exceeded those at 4 yr by up to 8 times.

Importance of characteristics of fire

Fires are variable, and a species of plant or animal that experiences high mortality in one fire may survive well in another, not because of any difference in the organisms, but because the fires are almost certain to differ. Variation in *fire intensity* provides clear examples of this. Crown fires can kill large trees, whereas low-intensity fires confined to the understorey cause little mortality. Different *fire types* can also produce different effects – a slow-moving, cool fire that burns the organic soil down to bedrock may cause greater tree mortality than an intense crown fire (Wade *et al.* 1980). A characteristic such as thick bark, which we might interpret to be of great survival value because it protects the plant in an intense but fast-moving fire, may actually increase susceptibility to a less intense but slow-moving fire, both because the bark ignites and perhaps because it retards the rate of cooling once the cambium has heated.

Varying fire intensity, rate of spread and continuity of the fire front may produce a variety of responses of animals to fires. The famous Florida naturalist, Archie Carr, told of occasions as a boy in north-central Florida when local landowners and volunteer fire fighters lined up along roadways in the face of an oncoming wildfire preparing to shoot mammals running before the flames, because those animals with smouldering fur acted as 'firebrands,' re-igniting the fire across the road. This picture contrasts strongly with the reports of mammals displaying no panic and remaining calmly within a home range while the fire front passed. The different observations may be explained by differences in fire intensity and rate of spread. For mobile animals caught in a fire, the speed and the uniformity of the fire front will determine the likelihood of escape, with greater mortality being caused by a fast-moving, continuous fire-front.

Other components of fire regime may produce equally variable responses. *Season of burning*, for example, is likely to influence mortality of both animals and plants, for a given fire intensity in a particular community. Mortality of longleaf pine trees is apparently greater after spring fires than after late summer fires (Cary 1932, Robbins and Myers 1992). In contrast, Cable (1973) recorded higher mortality of creosote-bush (*Larrea tridentata*) after a summer fire than after a winter fire in semi-desert of south-western North America. Different fire intensities and durations, interacting with the physiological state of the plant induced by season, may explain apparent contradictions such as these. For animals, season of burning may determine whether an animal is in a fire-tolerant stage of its life cycle (Robbins & Myers 1992, Stoddard 1935, Komarek 1965). Hunter (1905) (cited in Gandar 1982), for example, reported that grasshopper mortality is greater in a fire burning in cool, windy weather, because the fire front is moving rapidly but the insects are inactive and therefore unable to respond rapidly.

The *extent* of a fire is also likely to influence survival of animals, in particular. The continuity of the flame front and the frequency of occurrence of patches of unburned vegetation, reflecting continuity in the fuel and evenness in topography, will determine the ability of animals to dodge through flames and to find refuges. Thus, a grass fire may be more lethal to many animals, both invertebrate and vertebrate, than a forest fire, even though total fuel loads are almost inevitably greater in the latter.

Evolutionary responses to fire

It is common for evolutionary explanations to be offered for the observed responses of organisms to fire. The dangers in this approach were indicated at the outset of this chapter, and have been pointed out clearly by both Gould and Lewontin (1979) and Harper (1982). How can we sensibly come to ultimate explanations for observed characteristics? One main line of approach was argued by Harper (1982) as follows: 'The detailed analysis of proximal ecological events is the only means by which we can reasonably hope to inform our guesses about the ultimate causes of the ways in which organisms behave'. This approach will inevitably be slow in producing generalizations, because of the time and effort required to conduct detailed analyses of proximal ecological events, especially given the variability in natural ecological systems that can easily obscure patterns in short-term studies.

Grime (1985) argued that comparative approaches may also be used to advance our understanding of the evolutionary significance of particular traits. In particular, studies involving large numbers of species provide many opportunities for comparison and interpretation. Comparisons of traits characterizing successful and less-successful species in a habitat may suggest likely selective forces. Comparisons will undoubtedly reveal many differences and it will be difficult to determine which, if any, are of ecological significance. However, as more species are considered, it becomes progressively more difficult to fit more than one adaptive explanation to the data. Examination of populations of the same species, drawn from contrasted habitats will add to these inter-specific comparisons.

These various analyses of patterns proposed by Grime can be combined with the detailed population biology approach advocated by Harper. The analyses of pattern can identify the particular areas in which more detailed investigation will be valuable. In fact, there are several studies in fire ecology that have combined elements of these approaches (e.g. Keeley and Zedler 1978, Lamont 1985, Whelan 1985, Zammit and Westoby 1987, 1988, Cowling et al. 1987).

Many authors have made passing comments about the adaptive value of certain characteristics of plants and animals but there are relatively few studies that address this topic directly. The following discussion focusses on several characteristics that do appear to have adaptive significance in relation to fire and that have received some attention in the literature.

Sprouting versus obligate seeding in woody perennial shrubs

The co-occurrence of species with two sharply different modes of recovery after fire, sprouters and obligate seeders, has generated much interest. The lignotuber or 'burl' (James 1984), which provides the opportunity for rapid sprouting after fire, is so obvious after fire in many fire-prone regions that it has often been considered an adaptation to fire (e.g. Lopez-Soria and Castell 1992). Keeley and Zedler (1978) argued that sprouting after fire is an ancestral characteristic (see Wells 1969) and obligate seeding should therefore be viewed as the more specialized trait that has evolved in response to some fire regime. It has been argued that sprouting from roots and the presence of a subterranean storage organ permit survival of drought, frost, wind and herbivore damage to above-ground parts (see Bell *et al.* 1984). What is the adaptive significance of obligate seeding? Wells (1969) argued that the loss of the capacity to sprout after fire means that a species produces a greater number of sexually produced generations and there are thus regular opportunities for natural selection to be significant. Thus, an obligate seeding species might be expected to track a changing environment more successfully than a sprouter species. It is unclear why evolutionary tracking of relatively short-term environmental changes would in itself explain the success of an obligate seeder strategy, especially when many sprouter species also release sexually-produced cohorts with each fire, as well as surviving as established plants.

Inferring the importance of past fires in selecting for traits such as sprouting from rootstocks and obligate seeding requires regional comparisons, correlating what is known of fire history with the occurrence of these traits. In parallel with such a comparison, detailed population studies are required to assess the current adaptive value of the traits in relation to experimentally imposed fire regimes. Congeneric species-pairs would provide a good basis for comparison, and they exist in chaparral and coastal sage shrub in California (e.g. *Ceanothus* spp. and *Arctostaphylos* spp.; Keeley 1981), South Africa (e.g. *Leucospermum* spp.; Lamont 1985) and Australia (*Banksia* spp.; George 1984, Zammit and

Westoby 1988). Keeley and Zedler (1978) presented the testable hypothesis that a fire regime of short between-fire intervals would favour sprouting species while a fire following a long fire-free period would favour the obligate seeders.

Once a plant is committed as a sprouter or an obligate seeder, a variety of other traits may be adaptive under particular fire regimes. These include reproductive effort, breeding system, degree of protection of seeds and intensity of bradyspory. For reproductive effort, it is argued that a plant that does not devote resources to establishing and maintaining a large lignotuber should direct more resources to flowering and seed production. The other three traits above are forms of bet-hedging in order to avoid placing all the genetic future in one basket, so to speak, such that a single crop failure in a future generation ends the line of descent (Westoby 1981).

Carpenter and Recher (1979) postulated that because the seed bank is all-important in relation to plant fitness for obligate seeders, the following characteristics should be associated with obligate seeding: 'Seeders should invest more energy in flower, nectar and seed production than resprouters; seeders should be more attractive to pollinators than resprouters; seeders should have a fail-safe device to ensure some seed set in years when pollinators are few and/or are foraging in a sedentary manner and should therefore be more self-compatible than resprouters.' (Fulton and Carpenter 1979). Using similar reasoning, obligate seeders which are bradysporous would be expected to devote more resources than sprouters to protecting the stored seed bank from pre-dispersal damage by predators. It must be mentioned that the studies discussed in relation to this hypothesis have mostly examined only a single pair (sprouter *versus* obligate seeder) or at best a few species (Table 3.6), so none of them provides a good test of a putative, general relationship between sprouting and any of these other traits. Even with the limited sample of species examined, results are equivocal. In general, seedlings of obligate seeders appear to take less time to reach the age of first reproduction than seedlings of sprouters. However, as emphasized by Zammit and Westoby (1987), the sprouting stems of established plants may flower much sooner than the seedlings of co-occurring obligate seeders, and the significance of this for lifetime reproductive output is not yet clear. Studies by Carpenter (Fulton and Carpenter 1979, Carpenter and Recher 1979) suggest that obligate seeders may be more attractive to pollinators, be self compatible, and have a greater seed output per plant than sprouters. As a generalization, this is contradicted

Table 3.6. *Summary of studies comparing reproductive characteristics of co-occuring species of sprouters and obligate seeders*

	St John (1976)[a]	Fulton & Carpenter (1979)	Carpenter & Recher (1979)	Zammit & Westoby (1987)	Cowling et al. (1987)	Lamont (1985)
Obligate seeders	12 assorted Californian species	*Arctostaphylos pringelei*	*Banksia ericifolia*	*Banksia ericifolia*	*Banksia leptophylla & B. prionotes*	*Leucospermum cordifolium & L. erubescens*
Sprouters	3 species	*A. glandulosa var. mollis*	*B. spinulosa B. oblongifolia*[b]	*B. oblongifolia*	*B. attentuata & B. menziesii*	*L. cuneiforme*
Result	OS shorter time to 1st reproduction	OS more flowers/ plant, greater nectar production, higher nectar concentration, more pollinator visits, more selfing	OS more flowers per plant, more stored seed per plant, self-compatible. Sprouter breeding system unknown	OS shorter time to 1st reprod. for seedlings, longer cf. sprouting adults, seed prodn equiv. to sprouters for plants of same size	OS more seed produced per plant per year, more viable seed stored per plant (site 15 yr post-fire)	Sprouter more seeds per plant, sprouter self-compatible, sprouter more energy in nectar per flower head

Notes:
[a] Unpublished thesis; summary contained in Carpenter & Recher (1979)
[b] *B. oblongifolia = B. aspleniifolia* cited by Carpenter & Recher (1979); see George (1984).
OS = obligate seeders.

by the studies of Zammit and Westoby (1987), who found that an obligate seeder and sprouter species-pair had stored equal numbers of seeds when plants were matched for size in a site burned 15 yr previously. Lamont (1985) found that the one sprouter in a set of three *Leucospermum* species set more seeds per plant, produced more energy in nectar per flower head and was also self compatible. These few case-studies indicate that a wider survey is much needed.

Zammit and Westoby (1987) demonstrated that, for one sprouter/obligate seeder pair, germination after seed dispersal was highly synchronized for the sprouter and more spread out for the obligate seeder (Fig. 3.25). Timing of germination appeared to be under the control of the seed coat. The spreading of germination is interpreted as an adaptive trait, produced by selection operating on maternal plants, favouring those that spread germination of their offspring (seeds) over more than one period of desiccation. In a sprouter, lifetime fitness may be maximized by mass germination synchronized to the first rainfall event after post-fire dispersal. A single desiccation period may wipe out all the offspring after some fires, but this could be more than compensated by the great advantage conferred on those seedlings that are the first to appear after a fire, if the fire is followed by favourable climatic conditions. For an obligate seeder, this is a risky strategy, because a complete failure of germination after any one fire negates the fitness advantage gained from even the most successful germination event. Superimposed upon the variation in timing of seed germination described here for two bradysporous *Banksia* species, the timing of seed release from fruits in the canopy after fire can further 'spread the risk' of germination. Studies of this level of maternal control over the timing of germination needs to be added to studies such as those conducted by Zammit and Westoby.

The argument about risk-spreading can also be used to predict that obligate seeders would invest more in protection of a stored seed bank from predators and disease (Zammit and Westoby 1988). Thus, in the bradysporous species, seeds may be contained in tougher follicles and they may contain more toxic chemical defences than obligate seeders (Lamont 1993). This hypothesis receives some weak support from the study by Zammit and Westoby (1988), which showed that the proportion of the seed bank eaten by pre-dispersal predators was slightly less for an obligate seeder (*Banksia ericifolia*) than for a co-occurring sprouter species (*B. oblongifolia*). In contrast, Cowling *et al.* (1987) suggested that protection of seeds from predators is more likely simply to be linked with the intensity of bradyspory rather than to the mode of recovery

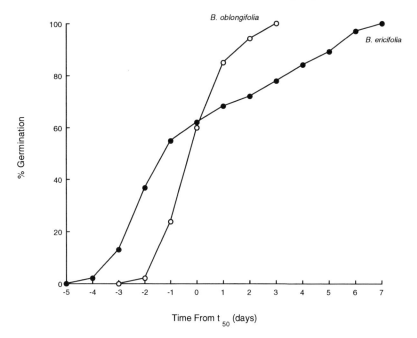

Figure 3.25   Comparison of the spread in timing of germination of seeds of two species of *Banksia*, one a sprouter (*B. oblongifolia* – open symbols) and the other an obligate seeder (*B. ericifolia* – solid symbols) (data from Zammit and Westoby 1987). Curves are standardized to the day on which 50% of the seeds had germinated (t_{50}).

after fire: protection of seeds from pre-dispersal predators should be a higher priority for individuals that store all of each seed crop for many years than for plants that release the seed crop annually.

Delayed versus immediate germination

One problem with the above argument about the traits that may be expected to be associated with sprouting or obligate seeding is that they are not all necessarily independent. Lamont *et al.* (1985), for example, provided information that permits a test of association between these traits for two genera, the South African *Leucadendron* species (collated by Williams 1972) and the Australian *Banksia* species (collated by George 1981). These classifications of species indicate a significant positive association between bradyspory and obligate seeding, at least in *Banksia* (Table 3.7).

It is difficult to interpret this sort of association and it is of course

Table 3.7. *Association between sprouting and delayed seed release (bradyspory) in two genera of Proteaceous shrubs: 85 species of Australian* Banksia[1] *and 81 species of African* Leucadendron[2]

	Number of species	
	Observed	Expected
Leucadendron		
Sprouters		
Bradysporous	4	3
Spontaneous release	2	3
Obligate seeders		
Bradysporous	37	38
Spontaneous release	38	37
$P = 0.24$; Fisher's exact test		
Banksia		
Sprouters		
Bradysporous	28	32.6
Spontaneous release	14	9.4
Obligate seeders		
Bradysporous	38	33.4
Spontaneous release	5	9.6
Chi-square $= 5.77$; $P < 0.05$		

Sources:
[1] Data from George (1984).
[2] Data from Williams (1972); cited by Lamont *et al.* (1985).

possible that both traits can be independently associated with a third causal factor, such as physical characteristics of the habitat – fire frequency, fire intensity and other forms of disturbance. This needs to be explored further. It would seem that bradyspory may be favoured in situations where the probability of a seed germinating and becoming established in any non-fire year exceeds the probability of it remaining viable during storage and subsequently becoming established after the next fire. There are two variables to be balanced here. It is easy to see how bradyspory could be favoured if all seeds stored on a plant retained viability for a long time and were well protected from seed predators, so that the next fire that provides opportunities for seedlings also releases most of the seed bank. However, bradyspory may also be favoured, even in an area with high pre-dispersal seed losses, if opportunities for germination and establishment are very strongly associated with fire.

If fires provide an environment more favourable for seed germination and seedling establishment than exists in unburned vegetation, fire *frequency* can readily be seen to be capable of influencing the success of bradyspory (Stern and Roche 1974). If fires are frequent, opportunities for successful seedling establishment may exceed the seed losses occurring during storage between fires (Whelan *et al*. in press). As the interval between consecutive fires increases, an increasing proportion of the pre-dispersal seed bank would be lost because of the combined effects of age, predators and disease and there would also be an increasing occurrence of conditions that are suitable for seedlings in the absence of fire. Givnish (1981) found a strong correlation between the proportion of bradysporous (called "serotinous' by Givnish) pitch pine trees (*Pinus rigida*) and the inferred fire frequency of the site in the pine barrens region of New Jersey. A causal role for high fire frequency in the evolution of bradyspory and the maintenance of bradysporous individuals within polymorphic populations is by no means certain. Lotan (1976) argued for high fire intensity, as well as frequency, as factors favouring closed-cone individuals of *Pinus contorta* (lodgepole pine). Cowling and Lamont (1985b) found a gradient in bradyspory in three Western Australian *Banksia* species associated with a climatic gradient. However, they identified several components of this climatic gradient, in addition to the inferred fire frequency, that may be associated with bradyspory. These other factors include the height of mature plants and moisture availability. It was argued that the height of a plant determines the intensity of the heat to which canopy-borne fruits are exposed. Fruits borne high in tall trees may not receive sufficient heat in a typical fire to cause seed release. Low moisture availability may determine post-dispersal survival of seeds by extending the time they spend on the soil surface awaiting suitable conditions for germination. Similarly, low moisture availability may limit seedling survivorship in between-fire years.

The selective advantage of bradyspory may lie in the synchronization of seed fall caused by fire. This synchronization of release of a large, accumulated seed bank has been shown to satiate populations of post-dispersal seed predators (Ashton 1979, O'Dowd and Gill 1984). In the case of the *Eucalyptus* species studied by O'Dowd and Gill, the principal seed-predators, ants, became more active after a fire but were nevertheless unable to cope with the vast increase in seed available on the soil surface. Seeds will presumably be accessible to surface-moving predators either until they are buried or until they germinate. The role of post-disperal seed predators in the ecology and evolution of bradyspory has yet to be examined in most systems. However, support for the potential importance of this interaction comes from studies on closed-cone pines in North America (e.g. Bramble and Goddard 1942) and on South African Proteaceae (e.g. Bond 1984) as well as the Australian studies mentioned above.

In any study of putative satiation, it is important to know whether the predators can 'catch up' with the immediate post-fire surplus before any seeds have escaped, either by germinating or being buried. Post-fire climate can have a marked impact here, because favourable post-fire conditions would minimize the time seeds spend on the soil surface, accessible to predators. There are few experimental tests of the predator satiation hypothesis. An interesting contrast with the findings of O'Dowd and Gill (1984) was described recently by Andersen (1988). In Andersen's study, post-fire seed fall from *Eucalyptus obliqua* trees persisted for several weeks. Much of the seed bank was therefore released when the seed-eating ants active on the forest floor were no longer satiated. This prolonged seed release was inferred to have been caused by a lower-intensity fire than that studied by O'Dowd and Gill (1984) in a *E. delegatensis* forest. Fire intensity could therefore determine, indirectly, the proportion of seeds escaping predators by regulating the amount of seed released and the synchrony of release. Ballardie and Whelan (1986) examined post-dispersal seed predation experimentally in mast-seeding and sparsely seeding populations of a cycad (*Macrozamia communis*). Fire is apparently one mechanism stimulating mass seed production in cycads (Burbidge and Whelan 1982, Beaton 1982). Contrary to expectation, seeds were more likely to be found and eaten by rodents in the masting site, in the midst of a great abundance of seeds. The contrast of these two examples with O'Dowd and Gill's study indicates that the detailed mechanisms and consequences of satiation of seed predators after fire need closer exploration.

Some species appear to hedge their bets, by displaying some degree of bradyspory. In many *Eucalyptus* species, for example, seeds are stored in capsules on the tree for several years before dehiscence. In this way, a fire will stimulate a mass release of several years of seed accumulation. If no fire has occurred during the first few years, the seeds are released anyway. By then, the risk of mortality during further storage on the plant may have exceeded the risk of failed germination and establishment following release. A different pattern is found in some of the closed-cone pine species, such as lodgepole pine (Lotan 1976) and jack pine (Gauthier *et al.* 1993), where an individual closed-cone plant may bear mostly open cones when it is young ($<$ 20–30 yr) or very old. The possible selective advantages of these characteristics remain to be investigated in detail, but Gauthier *et al.* (1993) argued that spontaneous opening of old cones may provide some insurance against very long fire intervals.

A comparison of the risk-of-storage and the risk-of-release is a good way to approach the question of the adaptive value of bradyspory. The risks associated with storage would include factors such as predation of stored seeds and loss of viability (e.g. Scott 1982, Bond 1985, Zammit and Hood 1986), and these would be weighed against the benefits of mass seed release once a fire occurred. Such benefits may include increased seed dispersal distances, predator satiation, improved conditions for germination and seedling establishment. The risks associated with release would include factors such as reduced seed dispersal, high post-dispersal seed predation and perhaps less favourable conditions for germination and seedling establishment. It is relatively simple to design experiments to determine the consequences of seed release between fires for a bradysporous species, measuring each of the above factors. With this approach, it may be possible to build a model predicting the optimal form of bradyspory under given climatic conditions and a particular fire regime for a plant with a particular life-history.

Timing of germination

In most fire-prone ecosystems, the timing of natural fires may vary over a period of several to many months. Even in Mediterranean-climate regions, natural fires (i.e. not of anthropogenic origin) may occur from early summer to late autumn (see Chapter 2). Where fires break seed dormancy, either by cracking an impervious seed-coat or by causing release of seeds from fruits stored in the canopy, there is some probability that the next rains that cause germination would be aseasonal and therefore give seedlings a 'false start'. Thus, Bond (1984) and Cowling and Lamont (1985a) have interpreted certain features of South African and Australian Proteaceae, respectively, as traits that delay germination until the season of the year when the likelihood of seedling establishment is greatest. After spring fires that caused seed release, nine of 11 Cape Proteaceae species studied by Bond exhibited delayed germination through the summer, although a few seeds of each species tested exhibited no dormancy and germinated immediately. Cowling and Lamont showed that dormancy in several Western Australian *Banksia* species was enforced by seeds being retained in the woody fruits, even after a spring fire had ruptured the follicles (see Wardrop 1983 and Gill 1976 for a description of the mechanism), until a series of wet and dry cycles caused the hygroscopic separator in each follicle to work its way out, carrying the seeds with it (Whelan 1986; Fig. 3.26).

If seed dormancy mechanisms such as these are indeed of adaptive significance, they would be expected to be strongest in seasonal environments, where the *predictability* of the season of good rainfall is high, and where natural fires typically occur either in a different season or with little seasonal predicability. Opportunities exist to compare the floras of different climatic zones as a test of the enforced dormancy hypothesis, even with sets of congeneric species. For example, the timing of both fires and rainfall is less predictable in the south-eastern United States coastal plain forests than in Californian chaparral. Similarly, there are summer- and winter-rainfall zones in both southern Africa and eastern Australia. Comparisons across these climatic regions, especially focussing on suites of congeneric species, would be very valuable in helping to infer the adaptive significance of some of the characteristics discussed above.

Flammability as an adaptive feature

The obvious flammability of many forest types across the world has led, from time-to-time, to speculation about whether flammability could be considered an adaptive trait. It has been argued that a species possessing characteristics that allow established plants to tolerate fires and/or ensure particularly good recruitment after fires would be even more strongly favoured if it was able to 'encourage' fires, thereby reducing the numbers of fire-sensitive competitors (Mutch 1970, Raup 1981, Cope and Chaloner 1985). This hypothesis has been debated (e.g. Buckley 1983, 1984, Snyder 1984) but few attempts at critical evaluation have been made. The problems with the argument that make the hypothesis difficult to test were outlined by Snyder (1984).

(i) Traits that permit recovery and/or recruitment after fire are widespread but flammability (for the whole community) is provided by only one or a few species (Snyder 1984).

(ii) It is easy to see other selective pathways that fortuitously cause high flammability (Snyder 1984, Cope and Chaloner 1985). For example, herbivory is widely recognized as a selective force that has resulted in the evolution of a wide array of secondary chemicals. Many of these are volatile and therefore flammable: the eucalypts of

Figure 3.26 Photograph of a *Banksia robur* infructescence showing many follicles (fruits) each enclosing a pair of winged seeds and the separator (arrow). As the humidity varies, the walls of the follicle open and close, as do the two layers of the separator. Together, these actions appear to work the seeds out of the fruit.

Australia and the chaparral shrubs of California are well-known examples.

(iii) It is difficult to see how an individual, high-flammability plant in a low-flammability population would be favoured, when fostering fire requires the 'cooperation of the whole community'. For example, some plant species in rainforests have high levels of oils and resins that make their leaves highly flammable in laboratory

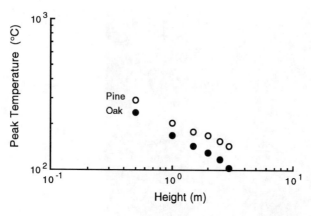

Figure 3.27 Temperature profiles beneath individual trees of longleaf pine (*Pinus palustris*) and a co-dominant tree species, turkey oak (*Quercus laevis*) in central Florida (data from Williamson and Black 1981). Temperatures under longleaf pine were considerably greater, at a given height, than those under oaks.

trials (King and Vines 1969), but the presence of such species does not ensure high fire frequencies or intensities in a rainforest.

Despite these arguments, the following case study suggests that flammability may be favoured as a result of the second-order interaction involving competition and fire. Many North American pine species can tolerate regular burning. In the south-eastern USA, pine stands are frequently subjected to invasion by a suite of fire-intolerant, hardwood species (Monk 1968, Veno 1976). The continuity of fuel under pine trees and high fire temperatures generated by pine needles appear to eliminate hardwoods without damaging the established pine trees (Williamson and Black 1981; Fig. 3.27). An individual pine tree that produced a less-flammable fuel would provide a cool spot in a fire, giving hardwoods more chance of establishment and long-term occupancy of the site than its own offspring.

Animal adaptations

The long list of plant characteristics that have been viewed as adaptations to fire makes it tempting to search for a similar list for animals. For example, Handley (1969) put forward a point of view that many find appealing (see also Allee *et al.* 1949):

Fire can be and often is a disaster for mammals dwelling in forests or other places where fires are infrequent. On the other hand, mammals living where fire is a regular and frequently occurring feature of the environment, as in grasslands, survive fires because of their adaptations to them.

This view of specific adaptation to fire is difficult to sustain, given that the catalogue of presumed adaptations includes mobility, burrowing, arboreal dwelling (Handley 1969) – all are common characteristics of mammals in many environments, including those rarely or never exposed to fires. Main (1981) concluded that, at least in Australian coastal heathlands, no animal species could be considered as having evolved tolerance of fire, except perhaps in relation to behaviour in the face of fire. In fire-prone regions, one feature of animals (especially mammals) faced with fire is a surprising lack of panic. This permits an animal to locate gaps in the fire front, to double back through the flames with least risk of injury, and to identify microsites that are least likely to burn intensely.

The history of entomology provides an intriguing possibility of fire specialists in the many references to 'fire-beetles', which appear to seek out recently burned vegetation for some reason, perhaps oviposition.

The beetle is known locally as the 'Fire-beetle' from its extraordinary habit... It is only seen when a bushfire is raging – in fact, the best way to take it is by starting one. The beetles seem to come from all quarters and fly straight into the fire, alighting and running about the hot steaming branches and sometimes even over the parts that are glowing red, yet without injury to the tarsi. (*H. M. Giles, reported by Poulton 1915*).

This habit is apparently not uncommon among beetles in the family Buprestidae, and anecdotes such as this one from western Australia also come from eastern Australia, the south-eastern USA, Canada and England (Poulton 1915). No doubt insects will provide a rich source of studies of putative fire adaptations.

Other evidence in support of fire specialists comes from species that have higher population densities some time after fire than before, or higher densities in burned than in unburned sites. This has been reported for lizards, birds and small mammals (e.g. Bock *et al.* 1976, Catling and Newsome 1981, Fox 1982). However, it is very difficult to determine whether these species depend entirely upon the regular occurrence of burned areas, or whether they are simply opportunists favoured by the change in vegetation and physical conditions after fire. The latter is probably the case. For example, species of the small rodents *Pseudomys* and *Mus* undergo enormous population expansion after fire and decline in abundance after several years. Can they therefore be viewed as "fire specialists"? However, similar population expansions of these species occur in the early stages of vegetation recovery after sand mining (Fox and Fox 1978). These animals can hardly be viewed as adapted to sand mining. Furthermore, *Mus musculus* is a cosmopolitan, opportunist species introduced into Australia relatively recently. A similar expansion of opportunist species is seen in the chaparral of western North America, where fire causes a shift in faunal species composition, temporarily favouring species of birds and mammals more typical of grassland (Lawrence 1966, Lillywhite 1977).

The only general conclusion that can be made at this time is that adaptation of animals to fire *per se* is very difficult to examine, either in relation to tolerance of the passage of fire or in relation to specialization on unique post-fire plant communities. It would be reasonable to expect any response of animals to fire to operate through the vegetation response, because the vegetation is such an important part of an animal's habitat.

Cryptic colouration

Melanism, especially in insects, is a trait which has contributed much to our understanding of predation as a selective force (Kettlewell 1973). Fire is a naturally occurring phenomenon that produces a black background at relatively frequent intervals, albeit for a short time-span. Moreover, predators such as raptorial birds are frequently seen congregating in recently burned areas. Increased predation and altered background colouration appear to be ingredients for a periodic intense selection for melanic forms of prey species, including mammals (Guthrie 1967, Kiltie 1989) and reptiles (Lillywhite et al. 1977), that survived the passage of the fire. This process is suggested by observations of colour change in arthropod populations in burned areas (Burtt 1951, Hocking 1964). Hocking's study, in the Sudan, showed that geophilous (grounddwelling) grasshopper species tended to be dark in colour while phytophilous species were mostly pale. Moreover, several species displayed intra-specific colour variations. It is not clear, however, whether paler individuals select paler backgrounds, or what cues are used to detect a suitable background. Also uncertain is the mechanism for a colour change within a population after fire. Simple selection by predators could account for it, as in the case of industrial melanisms. However, at a moult, formerly light-coloured nymphs could take on the darker shade of the burned background. Burtt (1951) demonstrated experimentally that adult grasshoppers (*Phorenula werneriana*) were able to take on the colour of various backgrounds, including burnt ground. Finally, as is the case for industrial melanism in moths, the actual advantage of being dark on a burnt background need not solely be escape from the attention of predators. Increased activity due to enchanced radiant heat absorption is also a possibility (Frost 1984) and there are several others (Jones 1982, Brakefield 1987). Much more work remains to be done on this interesting topic.

Fire-prone and fire-free environments

The above discussion indicates that various characteristics of organisms can be expected to have been influenced to some extent by past fires. However, among the published studies of the effects of fire, this disturbance is sometimes referred to as a catastrophic event, sometimes as a slight perturbation, and even as a necessary factor. These impressions flow from observations of mortality of individual organisms, alterations to population sizes or changes in community composition. There are many differences between ecosystems reported in the literature. Much attention in fire ecology has been focussed on the so-called fire-prone ecosystems, and the Mediterranean-climate regions of the world support well-known examples of these sorts of ecosystems. Here, fires are frequent, often started by natural events such as lightning, and organisms display various characteristics that permit individuals or populations to survive burning.

Much less work has been directed toward those environments in which fire is currently a very rare event, and has probably always been rare. It is these latter ecosystems that have produced the impression of fire as a catastrophe (*sensu* Harper 1977 p. 627). Harper distinguished between 'disasters' and 'catastrophes' as follows:

A disaster recurs frequently enough for there to be reasonable expectation of occurrence within the life cycles of successive generations – e.g. hurricanes at *ca* 70-year intervals are 'disasters' in the life of a forest tree and the selective consequences are likely to leave genetic and evolutionary memories in succeeding generations. A 'catastrophe' occurs sufficiently rarely that few of its selective consequences are relevant to the fitness of succeeding generations. The selective consequence of disasters is therefore to increase short-term fitness and the consequence of catastrophes is to decrease it.

The application of this explanation of the evolutionary effects of disturbances to understanding the effects of fire can easily become circular, if a fire-prone ecosystem is defined by characteristics of the organisms that might be viewed as fire adaptations. An examination of the evolutionary responses of organisms to fire first needs an independent estimate of the frequency of fires relative to the life-histories of the organisms under study. Much can be learned about the role of fire in the evolution of certain traits by a careful comparison of fire-prone and fire-free environments.

Outstanding questions

'The detailed analysis of proximal ecological events is the only means by which we can reasonably hope to inform our guesses about the ultimate causes of the ways in which organisms behave.' (Harper 1982). This quotation indicates one reason for focussing our attention on the detailed responses of individual organisms to fires. Detailed experimental studies will be crucial if we are to understand how organisms in fire-prone environments got to be the way they are, to understand the future evolutionary effects of fires, and to predict the effects of fires on populations and communities of organisms. Good experimental studies are one of the major challenges for fire ecology.

The following questions summarize some major gaps in our knowledge of the responses of individual organisms to fire.

1. What are the effects of season and frequency of fires on the mortality of established sprouter plant species?
2. What are the dynamics of seed banks of plants in fire-prone environments: both soil-stored and canopy-stored? Related to this, what conditions of fire and environmental characteristics favour the evolution of bradyspory?
3. What are the reasons for variation in the productivity responses of plants to fires – growth rates increasing as a result of fire in some cases but not in others?
4. What are seed dispersal distances in relation to spatial patterns of fires?
5. What are the responses of animals to intense wildfires, and how do these responses differ from those in low-intensity, prescription fires?
6. What roles do unburned patches of vegetation play in the survival of animals both during and after fires?
7. How does life-history influence survival of fire by animals, and how does this interact with the season of burning and likelihood of escape from a fire front?
8. What are the responses of organisms to fire in historically fire-free environments that do get burned?

4 · Approaches to population studies

This chapter and the following three will consider the broad topic of how populations and communities of organisms respond to fire. This is an important area of fire ecology, partly because of the large amount of research that is currently being conducted. There are many reasons for this. One is that management objectives focus attention on changes in population sizes of particular organisms and also on changes in community structure. In the last decade or so, there has been an increasing emphasis within ecology on population-level studies, particularly on studies of plant populations (e.g. Harper 1977, Silvertown 1982, Begon and Mortimer 1981, 1986). An understanding of population responses to fire is certainly important if evolutionary interpretations are to be made. Moreover, it will require detailed population studies to make predictions about the fates of particular species subjected to a changed fire regime. This is especially true for rare or endangered species.

'A population ecologist is interested in the numbers of a particular plant or animal to be found in an area and how and why population sizes change (or remain constant).' (Silvertown 1982; p. 2). What might be the aims of a population study in relation to fire? Of relevance to land management would be the ability to predict how a population or community will respond under the imposition of a certain fire regime. One may also wish to uncover the *mechanisms* underlying an observed change.

This brief chapter reviews approaches to the study of populations of plants and animals. Although much of this is fundamental, textbook ecology, it appears that fire ecologists have sometimes entered this field inadequately prepared for what is required of a sound experimental study. This is especially evident in the literature of animal population responses to fire. Any population study of the effects of fire would be improved by a prior reading of the principles of population ecology, found in a number of good texts (e.g. Varley *et al.* 1973, Harper 1977, Begon and Mortimer 1981, 1986, Silvertown 1982, Begon *et al.* 1990), and discussions of the need for adequate and appropriate forms of replication (Green 1979, Hurlbert 1984).

A discussion of various models of population growth and regulation is beyond the scope of this book and, in any case, many textbooks of ecology contain detailed treatments. However, it is valuable to consider various models of population growth and regulation for later comparisons with the likely patterns of population change in relation to fire. Fire could affect a population directly, through mortality, and indirectly, by altering other ecological factors that themselves operate on the population under study. These factors can include physical characteristics of the ecosystem and biotic interactions such as herbivory, predation, parasitism and competition.

Figure 4.1 categorizes the various interactions that may determine population size: density-dependent interactions *within* the population (i.e. intra-specific competition), density-dependent interactions with *external* factors (i.e. predation, herbivory, inter-specific competition), and non-regulatory influences (i.e. physical catastrophes, climatic extremes).

Figure 4.1 also illustrates various models of population dynamics that relate to these interactions. Part (A) represents logistic growth, in which a population grows exponentially until a limit is set by available resources. This asymptote, representing a balance between birth rate and death rate, emigration and immigration, is known as the 'carrying capacity'. Part (B) represents a more realistic expression of the logistic growth curve, in which the population fluctuates for a variety of reasons but its ceiling is ultimately set by the resources available in the habitat. Part (C) represents a population that fluctuates in size but whose ceiling is set below carrying capacity by some other density-dependent factor (e.g. predation). Part (D) represents a population so frequently decimated by disasters that population increase is always dominated by the exponential phase of the logistic curve. Finally, part (E) represents the dynamics of a population with sudden episodes of colonization (or recruitment from dormant propagules within the site) followed by population decline.

Which of these models applies to population change in relation to a fire regime? Fire is frequently depicted as an agent of disturbance that, in some areas, recurs with such a frequency that it could represent model (D): populations are always so affected by catastophes that population growth is always in the exponential phase. This model could explain the observed high species diversities of some fire-prone plant communities

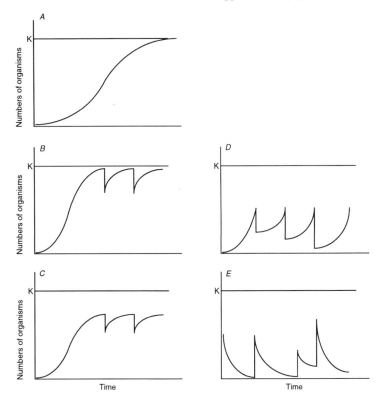

Figure 4.1 Models of population dynamics influenced by various factors
A & B. Density-dependent interactions within the population (e.g. intra-specific competition).
C. Density-dependent interactions with external factors (e.g. some forms of predation).
D & E. Non-regulatory influences (e.g. catastrophes, climatic extremes). *K* represents the population size at carrying capacity. Modified from Fig. 15.7 of Begon *et al.* (1986).

(see Cowling 1987) – frequent disturbance keeps population sizes low enough to prevent competitively inferior sprecies from being excluded (see also Chapter 7). However, much of the detailed research on the population responses to fires, particularly in regions viewed as fire-prone, suggests such a high degree of tolerance of fire by already established organisms that model (*E*) may be viewed as more appropriate.

Very few studies have censused populations frequently enough and/or for long enough to be able to identify the patterns of population change.

Detailed studies over shorter time spans, and measuring only some components of population dynamics – such as post-fire mortality, fecundity or recruitment – are more common

Assessment of population change caused by a fire

Mortality in a fire

Recording mortality after a fire is not, in itself, an indication that the fire has had a significant impact on the population. There are two main reasons for this. First, the individuals that died in the fire were not necessarily those contributing to population dynamics. If fire kills only weakened, sick, old or injured individuals that were destined to die even without the fire, there may be an alteration in the timing of their deaths but little or no effect on population-level fecundity. Second, any mortality caused directly by fire may be compensated for after some time by increased immigration, fecundity or survivorship of the remaining organisms. Additional information is needed before an effect of fire-caused mortality on a *population* is inferred – perhaps knowledge of the dynamics of both pre- and post-fire population sizes and knowledge of population dynamics in unburned, 'control' sites, for comparison with the burned site.

Estimates of population size

The basic tool of population ecology is a census; an estimate of the density of organisms in a sample area. Censuses have been used in a variety of ways in relation to fire but many of these approaches do not actually permit any inference to be made about the role of fire in producing the observed effect (Table 4.1). If any effort is to be invested in a study of the effects of fire, it is important to understand the limitations of the approach used and to be clear just what inferences may be made once results are obtained.

A simple and commonly used approach is to conduct censuses of the population before and after the fire. However, population sizes typically undergo marked annual and seasonal fluctuations even in the absence of fire, and other factors such as season, aseasonal climatic fluctuation and disease can cause rapid changes in population size. Any differences observed between before – and after – fire populations could therefore be adequately explained by these alternative factors, coincidentally

Table 4.1. *Various approaches that have been used to make inferences about the impact of a single fire on population size. This table indicates the appropriateness of the different approaches for each level of question asked, and points to possible shortcomings*

	Level of question asked			
	Indication of population change	Separation of fire effects from temporal change	Separation of fire effects from site effects	Inference of fire effect on population
Record of dead organisms after fire	No – dead individuals may not have contributed to population dynamics even without the fire	No	No	No
Census before and census after fire	Yes – but . . . choice of time for post-fire sample may mean a change is missed	No	No	No
Censuses before and after fire in a burned and a control site	Yes	Yes – but . . . confounding interaction of season & fire effects if timing of samples not carefully planned	No	No
Censuses before and after fires in randomly allocated replicate burned and control sites	Yes	Yes	Yes	Yes – but . . . care must be taken not to treat consecutive samples as replicates as they are not independent

occurring at about the same time as the fire. No amount of extension of censusing before or after fire can solve this problem.

Studies of fire have often focussed on wildfires, where neither the occurrence nor the location of the fire could have been anticipated or controlled. In this situation, pre-fire data are usually lacking, unless there has, fortuitously, been a previous study at the site. Comparison of the population of a burned site with that of a reference site that happened to escape the fire is therefore a common approach to population studies. This is not without serious limitations, because population densities vary in space, so any observed differences between the sites may have been there regardless of the fire. This is a fundamental flaw in statistical design of experiments, emphasized in 1930 by Fisher and Wishart! 'No one would now dream of testing the response to a treatment by comparing two plots, one treated and the other untreated' (Underwood 1986). Even with replication of samples within each site (burned and reference), it is possible only to conclude that the two *sites* differed in population density.

Replication of both sites that were burned and reference sites is an attempt to solve the limitations of interpretation outlined above. However, while replication will permit a conclusion about whether the burned sites are generally different from reference sites, there is still, inevitably, a confounding site effect that prevents the conclusion that the observed differences are *generally* attributable to fire. This is because a third, undetected factor may be controlling both the parameter under study (e.g. population density) and the occurrence of fires. Varying topography, for example, may determine the local population density of the organism under study, and also determine which sites burn in wildfires.

Ideally, replicate burned and control sites are needed, in which the fire treatment and the control treatment are allocated randomly to sites (Green 1979). This is an approach which should be used much more in future studies of fire ecology. It obviously requires that the fires be imposed as an experimental treatment. It seems strange, given the warning of Fisher and Wishart and the adoption of appropriate experimental approaches to studying the effects of fire on plant populations and communities in the 1940s (e.g. Wicht 1948), that there should be so few well-replicated and designed studies in fire ecology.

Understanding the effects of the experimental fire treatment against a background of seasonal changes in population size requires more than a single pre-fire sample and a single post-fire sample of the population.

Depending on the organism and its seasonal population fluctuations, studies should be initiated some time before the fire and continued for that long again afterwards. This has rarely been the case for fire studies. Even for arthropods, which are well suited to replicated fire studies, few studies have collected both pre- and post-fire population data for a full year each side of the fire (Table 4.2).

The experimental design of comparing a set of pre-fire censuses to a set of post-fire censuses within one area poses a number of difficulties for statistical analysis, particularly because lack of independence prevents the use of a temporal series of samples within one site as replicates (Hurlbert 1984). However, the experimental model of assessing the effects of fire on population size, by comparison of *control* and *treatment* sites *before* and *after* fire is analogous to some models of environmental impact assessment, and the discussions of appropriate experimental designs and statistical treatments by Green (1979), Hurlbert (1984) and Stewart-Oaten *et al.* (1986) are required reading for anyone embarking on a study of population responses to fire.

Having made a plea for the ideal of an experimental approach, with adequate and appropriate sorts of replication, it should be recognized that fires are not at all easy to deal with as an experimental manipulation. There are numerous reasons for this, but several come up especially often. First, the imposition and maintenenance of a fire regime for a specified, experimental block of vegetation is difficult. The weather will dictate whether conditions are too dangerous (e.g. dry and windy) to control an experimental fire at the set time. In other years, the site may be too wet to burn at all. Second, there is the problem of keeping out all unwanted fires. In a fire-prone environment, unburned treatments become more flammable each year. Third, there is a problem of scale in relation to replication.

The problem of scale is a serious one for reasons (only some of them logistical) associated with the area of vegetation that can be devoted to each replicate of each treatment. A simple randomized block design (e.g. Wicht 1948) could readily be applied, with each replicate of the order of 1–10 ha in some vegetation types. This design certainly replicates fires and unburned treatments, but if each fire covers such a small area, how relevant is the experiment to assessing the effects of wildfire? Herbivory presents an obvious problem. Some herbivores focus their attention on sites that have been recently burned (see Chapter 6). Small experimental fires fail to reduce herbivore populations and may serve to attract them when regeneration occurs. Statistically

Table 4.2. *Survey of 32 published papers describing the effects of fire on populations or communities of invertebrates*

Pre-fire data collected?		Long-term study (>1 yr)?		Both pre-fire and long-term data
Yes	No	Yes	No	
Abbott (1984a)	Andersen 1988	Abbott 1984a	Buffington 1967	Abbott 1984a
Hansen 1986	Athias-Bunche 1987	Andersen 1988	Cancelado & Yonke 1970	Majer 1984
Leonard 1974	Beckwith & Werner 1979	Athias-Binche 1987	Euler & Thompson 1978	Tester & Marshall 1961
Lilly & Hobbs 1962	Buffington 1967	Beckwith & Werner 1979	Evans et al. 1983	Wallace 1961
Lussenhop 1976	Cancelado & Yonke 1970	Hauge & Kvamme 1983	Gandar 1982	
Majer 1980	Euler & Thompson 1978	Jocque 1981	Hansen 1986	
O'Dowd & Gill 1984	Evans et al. 1983	Majer 1984	Haskins & Shaddy 1986	
Rice 1932	Gandar 1982	Merrett 1976	Izzara de 1977	
Riechert & Reeder 1972	Hauge & Kvamme 1983	Tamm 1986	Leonard 1974	
Tester & Marshall 1961	Izarra de 1977	Tester & Marshall 1961	Lilly & Hobbs 1962	
Wallace 1961	Jocque 1981	Wallace 1961	Lussenhop 1976	
	Merrett 1976		Majer 1980	
	Richardson & Holliday 1982		O'Dowd & Gill 1984	
	Rickard 1970		Rice 1932	
	Springett 1971		Richardson & Holliday 1982	
	Springett 1976		Rickard 1970	
	Springett 1979		Riechert & Reeder 1972	
	Strelein 1988		Springett 1971	
	Tamm 1986		Springett 1976	
	van Amburg et al. 1981		Springett 1979	
	Whelan et al. 1980		Strelein 1988	
			van Amburg et al. 1981	
			Whelan et al. 1980	

Source: From Tap (unpublished).

significant differences in such a study may simply reflect the effects of herbivory on artificially small patches of burned vegetation and therefore hold little relevance for wildfires. Including a fenced set of treatments doubles the size of the experiment and perhaps only introduces another artificial set of treatments (i.e. reduced herbivory). Expanding the size of each replicate plot so that it sensibly mimics the extent of a wildfire may be logistically impossible, and this approach would also ensure that replicate plots are initially very different from each other, necessarily being in different catchments, vegetation types, etc.

The combined effects of all these problems suggest that our knowledge of the effects of fires on populations and communities must increase through a variety of avenues: (i) unreplicated studies of wildfires that are otherwise well designed (i.e. astute use of unburned areas as reference sites) and well described so that future studies on another wildfire in the area may be treated as some sort of replication (see discussion on the uses of long-term ecological data by Westoby 1991); (ii) detailed studies of fires prescribed for various reasons (e.g. fuel-reduction and other management objectives), in which pre-fire data and data from sites to be left unburned can be collected but no replication of the fire treatment is possible; (iii) well-replicated, experimental studies, with care taken to incorporate the possible effects of scale in the interpretation.

Detection of long-term changes

The influence of fire on a population or community can be long-lasting, even though the disturbance itself passes very rapidly. Ecologists are often interested in post-fire changes in population size that may occur over the order of decades, but studies monitoring a single population for this length of time are rare indeed. The common solution to this problem is to conduct a *synchronic* study, also referred to as 'space-for-time substitution' (Pickett 1989). This can be defined as a study in which simultaneous measurements are made on a series of sites of different ages since disturbance, thought to represent a *chronosequence*. The best-known examples of this approach are studies of old-field successions (Billings 1938), glacial retreats (Crocker and Major 1955) and sand dune successions (Olson 1958).

There are serious limitations to the synchronic approach of estimating population change in relation to fire, and it is necessary to be well aware of the limitations before starting such a study. The main problem is the potential covariation of fire history differences and the observed population differences with a third (probably unknown) site-specific variable. For example, different population responses to a fire in adjacent sites last burned at different times in the past may indeed reflect an effect of time elapsed since burning. Alternatively, the factors that caused the differing fire frequencies (i.e. different climates, different productivities, unique topographic features, etc.) may independently be the cause of the observed population responses.

Bearing in mind the need for caution in the interpretation of a synchronic study, it is possible to obtain useful indications of the likely patterns of population change after fire. One of the earliest and most thorough examinations of the effects of fire on plants in any Australian plant community was by Specht and colleagues in the 1950s, studying the Dark Island Heath in South Australia (Specht *et al.* 1958). They recorded numbers of plants in six 5 × 10 yard quadrats in sites of varying post-fire age. Population sizes of seven dominant species showed a marked range of trends across the sites (Fig. 4.2*A*, *B*). Some species showed a consistent increase in population size over the age-sequence (e.g. *Phyllota pleurandroides* and *Banksia marginata*), some showed a decline over the age-sequence (e.g. *Banksia ornata* and *Casuarina pusilla*).

Within this set of data there is an indication that the set of sites may not have represented a good chronosequence. For example, the twofold difference in densities of *Leptospermum myrsinoides* between the 9-year site and both the 2.5 – and 15-year sites in Fig. 4.2*B* could simply reflect different fire histories at the different sites, rather than recruitment between 2.5 and 9 yr. Since this species produces seeds in abundance and also resprouts after fire, a previous fire followed by good germination and establishment may have started one site, 9 years previously, with a higher plant density than that initially present in either the 2.5- or the 15-year site.

In what amounts to a form of replication of the study for one species, Gill and McMahon (1986) found a very similar pattern for *Banksia ornata* to that reported by Specht *et al.* (1958) (Fig. 4.2*C*). Moreover, the general pattern of inferred survivorship in *B. ornata* is repeated in studies on other species with similar life-histories, for example *Eucalyptus regnans* (Fig. 4.2*D*).

If there is no alternative to the synchronic approach, careful selection of sites and appropriate study design can minimize the potential problems. For example, Auld (1984, 1986a,b, 1987) conducted a comprehensive study of the responses of populations of the legume *Acacia*

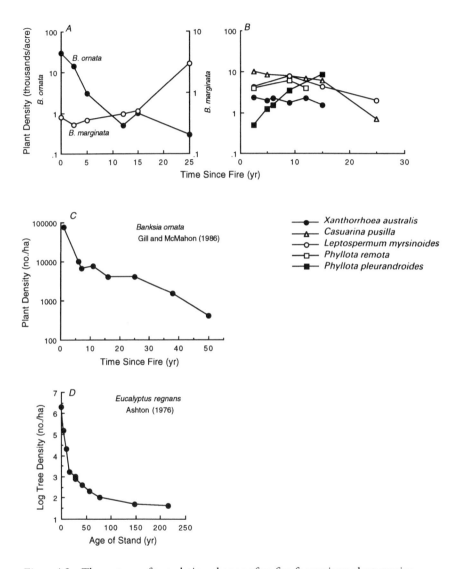

Figure 4.2 The pattern of population change after fire for various plant species, inferred from 'chronosequences' of sites of different post-fire ages.
A & B. Seven shrub and tree species from Dark Island Heath, South Australia (data from Specht *et al.* 1958).
C. Data from Gill and McMahon (1986) for *Banksia ornata*, supporting the result obtained in *A*.
D. Eucalyptus regnans (data from Ashton 1976).

suaveolens to fire in south-eastern Australian eucalypt woodland. His approach to the problem of the need for long-term studies was to examine a spatial set of plots representing different times since the last fire and to follow a tagged sample of plants in each plot for several years, thus combining synchronic and diachronic approaches (Fig 4.3). As long as some overlap in dates occurs between sites, there can be a check on the validity of the assumption that the spatial series actually represents a temporal series (see also Twigg *et al.* 1989 for a similar approach to studies of communities of small mammals).

Population change and fire regime

The effects of fires on populations could be assessed by correlating long-term changes in numbers with fire regimes. This 'diachronic' or longitudinal approach, if carried on for sufficient time with enough experimentally imposed treatments, could allow an assessment of the effects of each component of the fire regime.

The fact that there have been very few attempts at even this relatively simple study design for any population responding to a fire regime is a reflection of the long time scales required and the difficulty of imposing a constant fire regime as an experimental treatment. In the absence of studies of overall population changes through a series of fires upon which to base an assessment of the effects of particular fire regimes on a population, there is another approach that allows us to make predictions of population change from shorter studies. The effects of different fire regimes may be inferred from short-term studies of responses of various life-history stages to a single fire (see Gatsuk *et al.* 1980 for a discussion of this approach, and Wellington and Noble 1985a for a good example of its application). This approach is *diachronic*, because responses to different treatments are observed over time (Trabaud and Lepart 1981), but the overall population response is inferred from observations of the effects of fire on separate stages of the life-history.

Life-histories and the cohort approach

The simple technique of counting numbers of individuals to describe population change may obscure valuable explanatory information contained within the overall figures (Begon *et al.* 1986; Ch. 15). For example, an 8-year study of populations of sand dune plants in Poland (Symonides 1979) indicated that populations of some species declined

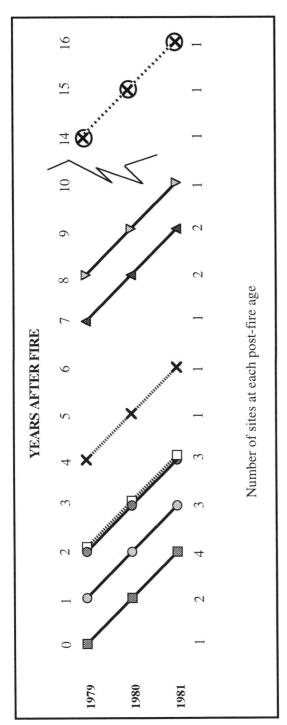

Figure 4.3 Design of the sampling by Auld (1986a, b) to provide support for the validity of a chronosequence by following each site for several years. Each symbol represents a different site sampled from 1979 to 1981.

while others increased. An explanation of *why* the observed, annual changes in population size were occurring was made more feasible by the splitting of the census into several life-history stages, revealing that annual variations in numbers of seeds germinating contributed substantially to the annual population fluctuations (see p. 113 of Silvertown 1982 for further analysis).

Begon and Mortimer (1981) provided a set of clear, diagrammatic representations of the life-history approach to the study of populations and this approach was developed for plant populations by Harper and White (1971) in a series of pictorial representations of plant life-histories

Data can be applied to these diagrammatic representations of life-history stages by censusing a particular *cohort* of individuals frequently enough through the lifespan. This may be easier for plant populations than for animals because, as succinctly put by Harper (1977), 'plants stand still to be counted and do not have to be trapped, shot, chased or estimated'. All that is required is the initial tagging effort, regular censuses and time. Even for plant populations, however, a complete cohort study is rarely conducted (for good examples, see Law 1979, Leverich and Levin 1979). For animal populations, emigration and immigration are likely to play important roles in determining the responses of populations to fire (much more important than the contribution these factors play in theoretical population models).

The graphical representation of population change in a cohort is a *survivorship curve* (Fig. 4.4). Studies of the effects of fire on recruitment to plant populations frequently employ survivorship curves to compare burned and unburned treatments. From such a graph, the effects of the treatment (fire) on lifespan, juvenile mortality rate and adult mortality rate can be clearly seen. As well as describing the temporal pattern of mortality, a cohort approach permits an assessment of age-specific demographic charateristics, such as reproductive rate.

The division of the life cycles of individual organisms into life-history stages provides the opportunity for predicting longer-term changes in populations due to an experimentally imposed fire regime on the strength of relatively short-term, diachronic studies. The dynamics of a cohort of each life-history stage can be followed for long enough to estimate transition to the next stage. This approach is becoming increasingly common in studies of responses of plant populations to specific fire regimes (see Chapter 5). The life-table approach to animal ecology has a long history, yet it does not appear to have significantly

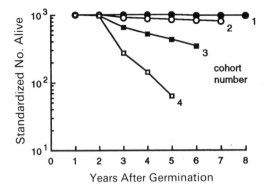

Figure 4.4 Examples of survivorship curves for *Agrostis curtisii* (modified from Gray 1987). The curves show the steadily declining chances of survival for seedlings in each successive cohort following a fire in 1976.

infiltrated studies of animal populations in relation to fire (see Chapter 6). It is important to remember that this approach makes the assumption, typical of all synchronic studies, that the processes operating now in one life-history stage will also apply to the previous cohorts when they eventually enter it.

Determining the causes of observed population change

Knowing how the components of a fire regime will influence birth, death, immigration and emigration rates may permit a prediction of population response to the fire regime. However, it has simply pushed the question of *why* the population has changed back to a set of questions about why birth, death and migration rates have changed. Fire can have primary effects on these factors, for example by causing mortality or altering fecundity directly, and it can have secondary effects, by altering other ecological processes that then affect population dynamics of the species in question. More complex attempts to model changes in populations must therefore focus on the ways in which birth rates, death rates and migration rates are affected by ecological factors such as population density (competition within the species), competition with other species, herbivory, predation and disease, as well as the primary effects of a fire (Silvertown 1982). Such studies will clearly require a manipulative, experimental approach.

Conclusion

The foregoing is designed as an introduction to population ecology in relation to fire, and a summary of problems of designing good population and community studies in relation to fire. What is already known, or can be inferred, about the responses of populations and communities to fires? The following three chapters examine the effects of fire on populations and communities of organisms; plant populations in Chapter 5, animal populations in Chapter 6 and communities in Chapter 7.

5 · *Plant populations*

Introduction

The simplest approach to population studies is to relate measured changes in total population size over time to the application of particular fire regime (see Chapter 4). This longitudinal (or *diachronic*; Trabaud and Lepart 1980) approach, if carried on for sufficient time with appropriate fire treatments, could allow an assessment of the effects of each component of a fire regime. There are only a few good examples of this approach in the fire literature. It will be clear from the shortcomings apparent in the examples and models described in this chapter that the establishment of long-term research sites subjected to a range of carefully applied fire regimes is urgently required.

Given the dearth of long-term studies of overall population changes through a series of fires upon which to base an assessment of the effects of particular fire regimes on any plant population, two main approaches allow the piecing together of tests of various population models (Chapter 4; see Fig. 4.1). Information about longer-term population changes after a fire may be obtained from *synchronic* studies of a series of sites thought to represent a chronosequence – an approach known as 'space-for-time substitution' (Pickett 1989). The effects of different fire regimes may also be inferred from short-term, diachronic studies of responses to a single fire of various life-history stages.

Explaining or predicting a population's response to a fire regime may be attempted by examining separately various life-history stages of the plant species and seeking information on the factors regulating three principal processes: (a) adult mortality caused by a fire; (b) density of germination after fire; and (c) establishment of seedlings and subsequent survivorship over time after fire. These processes are the focus of this chapter.

Adult mortality

Suites of terms such as *fire-sensitive decreasers, fire-resistant decreasers, and fire-resistant increasers* (Purdie 1977) or *invaders, evaders, avoiders, resisters, and endurers* (Rowe 1983), applied to plant population responses to fire, reflect the fact that different species respond differently (see Chapter 3; Table 3.4). This is true even within a single community subjected to one fire. The most conspicuous effect of fire is the mortality of adult plants. Despite some shortcomings with any classification system (see Chapter 3), it is possible to classify species according to whether or not established plants die in a fire. Thus, the terms *obligate seeder* and *sprouter* distinguish those species in which population recovery depends upon seeds because all adults are killed, from those species in which adult plants tolerate the passage of fire. It will be seen in the following discussion that divisions between these two categories are not always straightforward, because the ability to sprout may depend on characteristics of the fire. This observation has led to the definition of an intermediate category, *facultative sprouters* (see Chapter 3).

Obligate seeders

Obligate seeders provide a good illustration of variable responses tied to the characteristics of fire. Even within a single population, some plants may survive and others die due to different fire intensities at the location of each plant. For example, *Banksia ericifolia*, which is a tall shrub dominant in many eastern Australian woodlands and heaths, is widely held to be an obligate seeder with adults plants that are extremely fire-sensitive (Benson 1985). Nevertheless, individual plants that suffer only slight leaf scorch in a fire can sprout from stems (pers. obs.; Fig. 5.1). This variation in recovery as a result of different fire intensities led Gill (1981b) to suggest that a minimum fire intensity, defined as 100% leaf scorch, should be used in distinguishing a sprouter from an obligate seeder.

In addition to fire intensity, variations in other factors appear to influence survival of typical obligate seeders. *Banksia ericifolia*, for example, sprouted from rootstocks after heath fires that removed all above-ground biomass in coastal headlands at Jervis Bay, New South Wales (pers. obs. 1984, Fig. 5.1*B* Ingwersen 1977). Gill (1981b) suggested that genetic (ecotypic) variation, variation in plant stature and growth stage all combine to determine the likelihood of sprouting. A large-scale example of variation in sprouting is reported for *Purshia tridentata* (Bitter-bush) in the USA (Blaisdell and Mueggler 1956). Individuals of this species in California appear to be highly susceptible to fires, while widespread sprouting is reported from eastern Idaho.

Spatial patchiness of fires will determine what proportion of the adult plants in a population are actually exposed to sufficient heat to cause mortality. The role of topography in generating a mosaic of burned and unburned patches has been discussed in Chapter 2. In this context, categorizing a plant species by its sprouting response can be misleading in predicting a *population* response to fire. However, this effect of fire patchiness has rarely been examined directly in relation to plant population dynamics. Its importance lies mainly in the fact that survival of even a few adult plants, permitted by topographic or other discontinuities in the fuel (see McLoughlin and Bowers 1982, Benson 1985, Muston 1987), will produce a multi-aged population after fire rather than the single-aged regrowth stands that are frequently presumed. The importance of this effect on population dynamics will depend upon the relative scales of the patchiness of a fire and the distances of seed dispersal.

Selection of study sites for the assessment of fire effects on populations of species of obligate seeders is likely to have been influenced by a desire for uniformity of age- or size-classes both within and between treatments (e.g. different densities, different fire seasons; Morris and Myerscough 1988). Quantification of the proportions of adult individuals surviving within populations of obligate seeder species in various fires is needed, especially in relation to the assessment of the long-term impact of altering fire frequency. In sites where all plants die in each fire, one fire closely followed by another would indeed cause population extinction if the seed bank is depleted after the first fire, through germination, and if these seedlings have not reproduced by the second fire. However, as discussed previously, frequent fires are likely to be both less intense and also more patchy. How does this patchiness affect the survival of adults in each fire and the subsequent production of seedlings?

Sprouters

Some established plants of fire-sensitive species may survive a given fire. The other side of the coin is that there can be some mortality of fire-tolerant species. The fire type, intensity, frequency and season might all be expected to influence mortality rates. Wade *et al.* (1980) described a peat fire that caused mortality of adult pine and cypress trees, which are normally tolerant of even moderate- and high-intensity ground fires, by affecting root systems. Likewise, Cypert (1961) reported that few cypress trees survived in those patches of Okefenokee Swamp, USA, that sustained a peat fire.

Bell (1985) listed several species in Western Australian sandplains that

Figure 5.1   *A.* Stem sprouting in *Banksia ericifolia* in Hawkesbury Sandstone woodland near Wollongong, Australia. This woody perennial shrub species (Proteaceae), common in eastern Australia, is typically considered to be an obligate seeder.
B. Banksia ericifolia can also sprout from underground parts, as shown here after a fire in coastal heathland at Jervis Bay, eastern Australia.

display variable survival depending upon fire intensity. The effects of fire intensity are likely to interact with plant size in a population represented by many age- or size-classes. Burrows (1985), for example, showed that the proportion of *Banksia grandis* plants killed back to ground level was a function of fire intensity. A surprising finding in his study was that the probability of sprouting (i.e. coppicing) after having been killed back to ground level actually *decreased* with increasing stem diameter. Similar observations have been made of other plant species (Blaisdell and Mueggler 1956, Baird 1977, Williamson and Black 1981). From the results of these studies, it is clear that mortality in a population will depend on both fire intensity and also size- or age-distributions of plants prior to the fire.

The ability of frequent fires to cause rapid population change in sprouter species was well demonstrated in the Californian chaparral by Zedler *et al.* (1983) (see Fig. 2.18). One of the species in this study, *Xylococcus bicolor*, sustained only 0–5% mortality during the first fire but the second fire, a year later, caused between 17 and 31% mortality. Noble (1982, 1984, 1989) has established one of the few long-term, experimental studies of demography of a woody perennial plant in relation to fire. The mallee eucalypts typically sprout from epicormic and lignotuberous meristems after fire (Jacobs 1955). However, two successive wildfires in the New South Wales arid zone, in 1975 and 1977, killed approximately 20% of adult plants (Noble 1982). Consequently, a series of experimental fires was established in 1978 to compare mallee survivorship after fires in different seasons. Cereal straw (5000 kg/ha) was applied to the treatment blocks to permit the burning regime. Mortality rates were quite different under the two seasonal treatments (Fig. 5.2)

Significant mortality occurred after consecutive autumn fires, which were more intense than the spring fires despite the experimental control of fuel biomass. Annual fires would not usually occur naturally in this area because of the slow rate of fuel accumulation, but these data may nevertheless reflect a difference in the physiological status of plants in different seasons. Furthermore, these data provide a hint that autumn fires every 3 years may permit greater survival than either annual or biennial fires. Future trends in the results of this experiment will be very interesting, and setting up experiments like it in other ecosystems would be very valuable.

Season of burning and plant physiological status have been shown to interact to influence sprouting after defoliation of young plants in

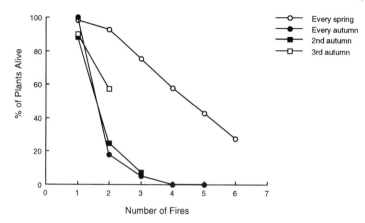

Figure 5.2 Survivorship in a sample of mallee plants (*Eucalyptus* spp.) subjected to four experimental fire regimes varying in both frequency and season (from Noble 1982). Spring fires, even every year, do not appear to cause as much mortality as any of the fire frequencies in autumn.

Australian eucalypt forests (Cremer 1973). Recovery after defoliation was poor after periods of rapid shoot growth and high after periods of quiescence. In comparisons of fast-growing and slow-growing plants occurring in sites of differing productivity, good recovery among fast-growing plants was confined to a much shorter period of the year (Fig. 5.3). Although these data refer to clipping experiments, they support the above observations made of the recovery of mallee populations after fires in different seasons. Similar observations have been made for a variety of species, including conifers. Cremer (1973) related this pattern of different degrees of sprouting in different seasons to the accumulation and storage of starch in both roots and stems (Fig. 5.4). In periods of rapid growth, starch is directed to growing shoots and little is available to support sprouting if defoliation occurs at such a time.

The effects of differing fire frequencies on mortality of established plants are related to the ability of these plants to continue replenishing a bud-bank, either epicormic or in a lignotuber. Zammit (1988) assessed the ability of a lignotuberous shrub, *Banksia oblongifolia*, to keep producing new vegetative buds by consecutive clippings of shoots (ramets) on 20 plants sprouting after a fire in May 1982. After four successive clippings over 18 months, no further sprouting was observed (Fig. 5.5*A*). These results presumably under-estimate losses from the bud bank due to fire, because the number of ramets produced per dm^2 was

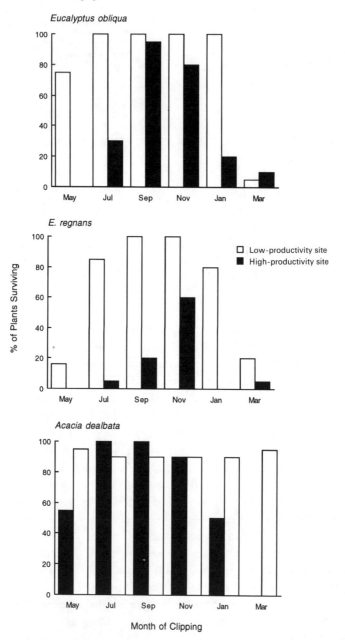

Figure 5.3 Recovery of three tree species after defoliation by clipping in different months for plants growing in sites of low productivity (open bars) and high productivity (solid bars) (from Cremer 1973).

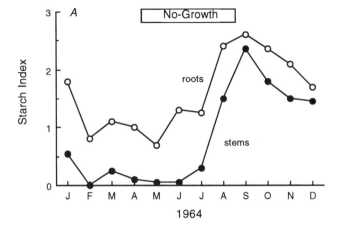

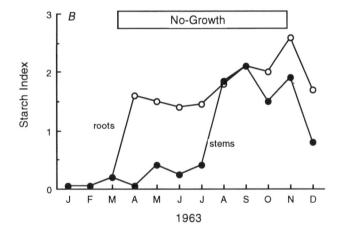

Figure 5.4   The store of starch in roots and stems varies over the year, being accumulated through the no-growth period. These results (from Cremer 1973) are for *Eucalyptus regnans* growing in sites of low productivity, sampled in 1964 (*A*), and high productivity, sampled in 1963 (*B*).

lower after fire than after clipping (Fig. 5.5*B*), implying that the heat of the fire killed some buds that would have sprouted if the plant had merely been clipped.

The capacity for repeated recovery after frequent fires is highly variable among species. In contrast to Zammit's study of *Banksia*, Chattaway (1958) found that *Eucalyptus* seedlings could be clipped of all leaves for up to 26 times before finally succumbing.

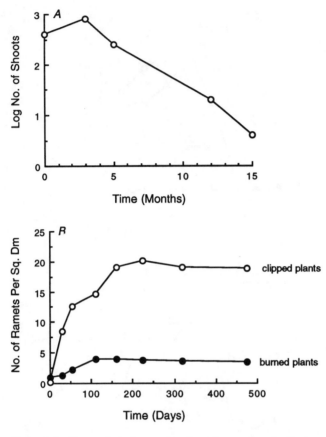

Figure 5.5 Dynamics of sprouting from lignotubers in *Banksia oblongifolia* in eastern Australia (data from Zammit 1988).

A. The bud-bank of meristems in lignotubers available for sprouting was inferred by counting shoots resulting from successive clippings, until no further shoots appeared. The datum at 0 months represents the standing crop of shoots on the sample of 20 plants at the time of first clipping. Subsequent data points represent the numbers of new shoots appearing by the clipping at that time. The clipping at 15 months produced no further shoots.

B. Comparison of the cumulative numbers of shoots appearing per dm² of root crown after either burning or clipping in May 1982.

Interactions of fire with other factors

The interaction of fire with other ecological factors has the potential to produce mortality where fire alone would have little impact. Herbivory and drought are two clear examples of this sort of interaction. Leigh and Holgate (1979) studied the effects of herbivory on survival of both seedlings and established plants after fire, for several sprouter species in

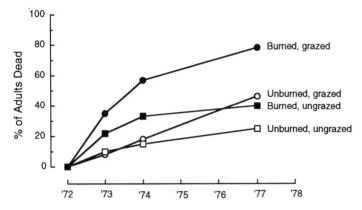

Figure 5.6 Experimental manipulation of herbivory (by caging) and fire indicates the importance of the interaction of these two factors in survival of the otherwise long-lived perennial legume, *Daviesia mimosoides* (data from Leigh and Holgate 1979). Solid symbols represent burned plants, open symbols – unburned. Circles represent grazed plants (i.e. uncaged), squares – ungrazed.

Australian eucalypt woodland. *Daviesia mimosoides* (Fig. 5.6) is a good illustration of the potential impact of herbivory on adult plants – mortality in plots that were either burned and not grazed or grazed and not burned was only slightly greater than that recorded on the control sites, which were neither burned nor grazed, but the combination of burning and post-fire grazing produced double the mortality rate for the burned-only or grazed-only treatments.

This interaction may be much more important than it would appear from cursory observations of a burned site. The true impact of herbivory on a plant population may be revealed only in the sort of exclosure experiment performed by Leigh and Holgate. If the magnitude of the effects detected in this study is general, then the impact of fire on herbivore populations becomes of great significance in plant population and community studies. In particular, the size of a burned area used in experimental studies will be significant, not only because plants may be particularly susceptible to herbivory as seedlings or sprouts, but especially because herbivores are likely to congregate on patches of burned vegetation.

Density of germination

Recruitment after fire is dependent upon several factors relating to the seed bank: (i) size of the dormant seed bank accumulated since the last fire; (ii) mortality of seeds in the fire; (iii) proportion of seeds released

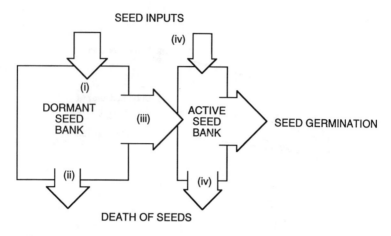

SEED INPUTS

Figure 5.7 Model of seed bank dynamics. The potential for recruitment after fire depends on many factors: (i) size of the dormant seed bank accumulated since the last fire (ii) mortality of seeds in the fire (iii) proportion of seeds released from dormancy by the fire (= germination) and (iv) postfire production (and losses) of non-dormant seeds.

from dormancy by the fire (= germination); and (iv) post-fire production (and losses) of non-dormant seeds. These processes are summarized in Fig. 5.7.

Reproduction

Although a substantial amount of information is now available describing flowering in a variety of plant species after fire, little attempt appears to have been made to relate these observations to quantitative measures of population dynamics. Age at first reproduction is a fundamental feature of the innate capacity for increase of a population, r_m, because it is directly related to generation length. The following equation describes the relationship:

$$r_m = \frac{\log_e(R_0)}{G}$$

Where R_0 = net reproductive rate; and G = the mean generation length (see Ch. 11 of Krebs 1985).

Age at first reproduction varies greatly among species within a single community (see Fig. 5.8A), between populations of a single species and also after different fires within the one population. It could therefore be an explanation for differences in population responses to fires.

A

	Years to first flowering in SITE:-				
	A	B	C	D	E
Actinotus minor	2	-	-	3	-
Gonocarpus teucroides	2	-	4	-	-
Boronia ledifolia	4	-	-	5	-
Gompholobium grandiflorum	4	-	-	6	-
Eriostemon buxifolius	-	-	5	5	4
Acacia suaveolens	-	-	-	5	4
Hakea teretifolia	6	6	-	7	-
Petrophile pulchella	-	6	-	9	8
Banksia ericifolia	8	8	-	7	-

B

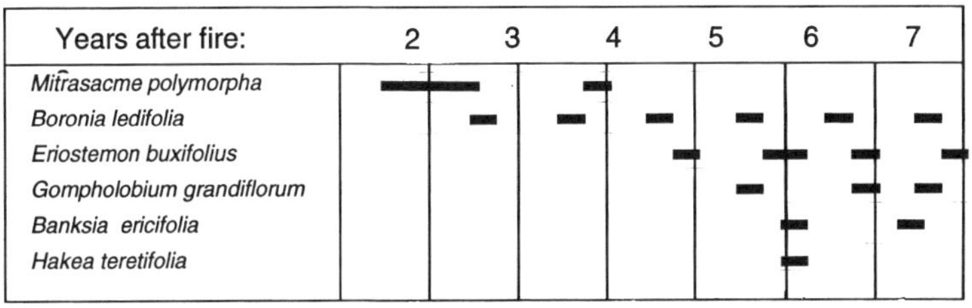

Years after fire:	2	3	4	5	6	7
Mitrasacme polymorpha						
Boronia ledifolia						
Eriostemon buxifolius						
Gompholobium grandiflorum						
Banksia ericifolia						
Hakea teretifolia						

Figure 5.8 Life-history characteristics of plants in relation to fire can be quite variable, as illustrated in a study by Benson (1985) in sandstone woodland and heath vegetation near Sydney, Australia.
A. Time to first flowering of a variety of woody perennial shrub species varied among sites.
B. Year of first flowering varied among species.

The simplest studies of the effect of fire on flowering and seed production would be conducted on populations of obligate seeder species with a single cohort of post-fire recruits producing an even-aged stand; i.e. a population with non-overlapping generations. Flowering in a cohort clearly varies with age. Seedlings must reach a certain age or size before first reproduction. *Banksia ericifolia* is a woody perennial shrub common in heath and woodland communities of eastern Australia. Observations on single cohorts suggest that time to first flowering is about 6 years (Fig. 5.8*B*). However, first flowering can occur in the third year in favourable sites, and may be delayed for 9 years or more in less-favourable sites (Muston 1987). Observations of this first flowering in *B. ericifolia*, in whichever post-fire year it occurs, suggest that seed production is very poor (Bradstock and O'Connell 1988). Benson (1985) reported a similar finding for another member of the Proteaceae, *Grevillea sericea*. Possible reasons include poor pollination when overall flower densities are low (i.e. few inflorescences per plant) or inadequate

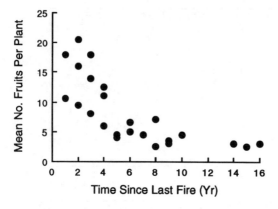

Figure 5.9 Decline in the mean number of fruits per fruiting plant of *Acacia suaveolens* over time since the last fire (data from Auld and Myerscough 1986).

stored resources in small plants to permit seed production even after pollination and fertilization.

The intensity of flowering may reach a peak and then decline with age. For example, in a study of the short-lived, obligate-seeding legume, *Acacia suaveolens*, Auld and Myerscough (1986) found that the mean number of fruits per fruiting plant declined sharply with age (Fig. 5.9), mostly because of shoot dieback in older plants. The proportion of plants producing fruits did not vary consistently with the age of the cohort. Development of fruits is apparently confined to those shoots produced the previous year and shoot production declines as the plants age. These results show that, within a generation, the greatest contribution to fecundity is made in the first few years after germination. Whether the greatest contribution to *recruitment* also comes from these years of peak reproduction depends on many other factors, especially the mortality rate of seeds in a soil- or plant-stored seed bank (see below).

This same pattern of no flowering in the seedling stage, followed by a peak and then a decline with age, may hold for a variety of other species, although on different time scales (see, for example, the very rapid decline in flowering in two Western Australian kangaroo-paw (Haemodoraceae) species within 5 years of fire; Lamont and Runciman 1993). However, too little information is available to permit generalization. Little decline in flowering per plant has been detected for long-lived shrubs, in which flowering continues to increase as the plants increase in size. For example, studies of the obligate seeder *Banksia ornata* (Proteaceae) show a continuing increase in flowering with age of stand until old

Table 5.1. *Flowering intensities and fruit production of* Banksia grandis *after a moderate-intensity fire in Western Australian eucalypt forest*

		1980	1981	1982
No. of trees in flower	Burned	0	0	6
	Unburned	11	20	24
Total no. inflorescences	Burned	0	0	9
	Unburned	29	54	143
Total no. fruiting cones	Burned	0	0	0
	Unburned	2	4	10^a

Note:
[a] estimated from the proportion (0.07) of cones to inflorescences recorded in 1980 and 1981.
Source: After Abbott (1985).

plants suffer dieback and death after about 50 yr (Specht *et al.* 1958, Gill and McMahon 1986).

Similar trends occur in the flowering of sprouting plants after fire. In the seedling generation there may be a substantial period before first flowering, but this remains to be quantified for most species. Benson (1985) found that seedlings of some sprouting legume species in eastern Australian eucalypt forest and heath produced flowers within 3 years of germination (Fig. 5.8), whereas seedlings of other species had not reached this stage by 10 years . Perhaps there are broad patterns among species in this characteristic (i.e. do sprouting and obligate-seeding species differ consistently?).

In the adult generation of a population of a sprouting species, there may also be a period without flowering after fire, while resources are devoted to vegetative growth. This delay in flowering and seed production as a result of fire has rarely been quantified. However, Abbott (1985) reported that *Banksia grandis*, an abundant tree species in south-western Australian eucalypt forests, failed to produce inflorescences in the first two years following a moderate-intensity fire, and limited flowering in the third year failed to produce fruit (Table 5.1).

The factors that affect post-fire sprouting are also likely to affect post-fire flowering when the numbers of flowers or inflorescences per plant are determined by the number of active shoots, as was the case for *Acacia suaveolens* (above). An example of this is the Waratah (*Telopea speciosissima*; Proteaceae), an eastern Australian shrub. Each shoot sprouting after

Table 5.2. *Flowering of* Banksia paludosa *in eastern Australian sedgeland in relation to time since last fire. Site 1 was burned in 1980 and again in 1983 – hence only 3 years of data were available. Site 2 was also burned in the 1980 fire but was not surveyed until 1984*

		Years after 1980 wildfire			
		1	2	3	4
% of sample	Site 1, $n = 59$	0	5.1	66.1	a
flowering	Site 2, $n = 29$	a	a	a	55.2
Mean no. (n) in-	Site 1	0	2.5 (2)	5.5 (39)	a
florescences per	Site 2	a	a	a	4.0 (16)
non-barren plant					

Notes:
Numbers in brackets are sample sizes.
a data not collected.
Source: R. J. Whelan unpublished data.

fire can produce a terminal inflorescence. Flowering in garden plants of this species can be enhanced by pruning (Wrigley and Fagg 1979), which stimulates the production of new shoots. A fire disturbance that maximizes sprouting from lignotuber meristems may therefore be expected to enhance post-fire flowering.

In addition to a delay in the build-up of flowering after fire, some studies have suggested that plants in sites that have escaped fire for a long time also show poor flowering. For example, in sprouting *Banksia* shrubs, the pre-reproductive period may be 3 years or so (Table 5.2) and flowering in plants unburned for 14 years may be poorer than that on plants last burned 5 years previously. Pyke (1983) obtained similar flowering data for two other Proteaceae species. Although data such as these do not adequately exclude the possibility of a site-effect, rather than reflecting the effect of time since last fire, they do suggest that the maximum contribution to fecundity may occur in the intermediate years after fire.

Fire can produce a marked stimulus to flowering of sprouting herbs and shrubs (Le Maitre and Brown 1992). The most spectacular examples, as discussed in Chapter 3, are those species that seem to flower only (or predominantly) in response to a fire, such as some orchids (Erickson 1965), grasses (Daubenmire 1968) and some species of grass trees (e.g. *Xanthorrhoea australis*; Gill and Ingwersen 1976, Gill 1981b) in Australia. The effects of season or intensity of fire on flowering intensity in

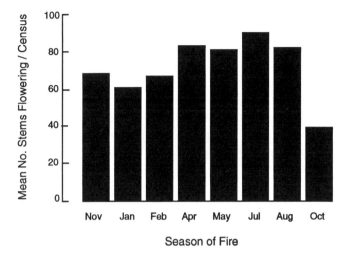

Figure 5.10 Mean numbers of stems flowering per census for one year after experimental fires that were set at seven different times of year. Sites were longleaf pine (*Pinus palustris*) flatwoods in northern Florida (data from Platt *et al.* 1988).

sprouting species have rarely been quantified. However, Le Maitre and Brown (1992) found that *Watsonia borbonica* (Iridaceae), a geophyte in South African mountain fynbos, flowered much better after summer or autumn fires (50–80% of ramets) than after spring or winter fires (< 10% of ramets). Many species of herbaceous perennial sprouters in Florida longleaf pine communities flower better in the year after fire than in sites unburned for several years (Whelan 1985). Platt *et al.* (1988), in a rare experimental study of flowering in sites burned in different months of the year, discovered several effects (Fig. 5.10). Fires during the growing season synchronized flowering within populations and between species and also produced shorter flowering spans than fires between growing seasons. Plants burned in the growing season also produced more flowering stems. It is only possible to speculate of the impact of these effects on subsequent population dynamics. Higher levels of seed set might be expected to result from synchronized flowering within a population, as pollinator visits and outcrossing may be facilitated. However, overlap in flowering between species may produce more inter-specific, and therefore ineffective, pollinations. Mass flowering might also satiate polllinator populations, though this possibility does not appear to have been examined thoroughly.

Synchronous flowering stimulated by fire is linked with the phenomenon of mast fruiting, in which a season of abundant seed production is followed by several seasons of poor reproduction. Although there have been few experimental tests of the consequences of this phenomenon for a plant population, satiation of seed predators is a favoured explanation (Silvertown 1982, Ballardie and Whelan 1986). The consequences of mast fruiting are discussed more fully below.

The timing of a fire in relation to a normal mast-flowering cycle is of potentially great importance to plant population dynamics. A fire in the year of abundant flowering could remove, in the bud stage, the chances of reproduction for perhaps a decade. In contrast, a fire just after seed production in a mast year may enhance recruitment to the population, for various reasons associated with seed survival and seedling establishment discussed below. For species that do not display a mast-flowering cycle, the season of burning may produce similar effects. Burning during flower initiation and development can remove a full year's potential recruitment.

Seed-bank dynamics

Although very little quantitative work has focussed on the changes to seed banks in relation to fire, they are potentially of great importance in population dynamics. Cavers (1983) stated that: 'the fate of a plant population may be decided by the pattern of mortality exhibited by its seeds. Further, the results of the interactions between populations of different species may be decided by the relative mortality patterns of the seeds of these populations, as has been shown by Grubb et al. (1982).' In view of the fact that species in fire-prone plant communities usually display a variety of patterns of seed release, seed retention and seed dormancy, and also in view of the obvious effects (in terms of both mortality and release from innate dormancy) that fires have on seeds, it is surprising that so few studies have been conducted on seed banks. From an evolutionary point of view, the immense mortality that occurs in the seed stage (i.e. 'at least 95% of all plant mortality'; Hickman 1979), if it is non-random with respect to genotype, provides potential for marked evolutionary change within a population. This may be especially true for species in which established plants die in a fire and population persistence occurs solely through a seed bank (Gray 1987).

The major factor determining the development of a perennial seed bank is seed dormancy. Most fire-prone plant communities contain both species that possess innate seed dormancy (e.g. impermeable seed coat cracked by the heat of fire; seeds retained in tightly closed fruiting structures stimulated to open by fire; see Whelan 1986) and also species that produce seeds that will germinate readily in the appropriate conditions of oxygen, heat and moisture. This latter group may include species with no dormancy at all, in which ungerminated seeds soon die, and also species in which viable, germinable seeds are stored in locations where conditions enforce dormancy. Because the consequences of fire for 'populations' of seeds depend on the sort of dormancy mechanisms displayed and also on the site of storage of dormant seeds, we discuss these categories separately below.

Species with no dormancy
Many plant species do not develop a long-lived seed bank, most seeds either germinating as soon as they are released and soil conditions become appropriate, or dying for one of various reasons, including seed predators and decomposition. Potential annual inputs to the population therefore depend directly upon annual seed production by established plants. Hassan and West (1986) studied the soil-stored seed banks for one full year in burnt and unburnt sites of a sagebrush community in Utah, USA. In this community, the seed bank of most species had virtually disappeared, even in unburned sites, by mid-summer, diminished by germination, decomposition and granivory. Thus, there was very little or no carry-over of seeds from one year to the next. A wildfire in mid-summer (July 1981) would have burned many populations during flowering or seed production and therefore produced a decline in total stored, viable seeds of about 50% by September (Fig. 5.11). For most of the native species, fire also retarded the recovery of the seed bank the next year (i.e. by September 1982), following the June nadir. This presumably resulted from mortality of established plants and/or reduced flowering of the survivors. However, seed dispersal from adjacent, unburnt plants may have contributed to the seed banks present in the burned samples by September.

The introduced annual grass *Bromus tectorum* showed an increased seed bank in the burned sites by September 1982, in contrast to the native species. The reason for this is not known, but plants resulting from the germination of those seeds surviving the fire in July 1981 may have reproduced better than their unburned counterparts and perhaps seed losses in autumn 1982 (i.e. to granivores) were lower in burned than in unburned sites. This study illustrates the importance of following seed banks for several seasons in both burned and unburned sites. It also

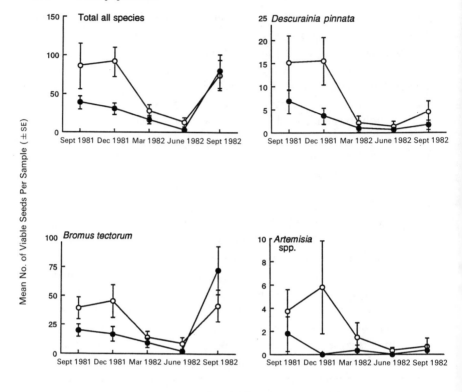

Figure 5.11 Variation over a year in soil-stored seed banks of several plant taxa in a sagebrush community in Utah, USA (data from Hassan and West 1986). Samples were taken from pairs of plots (plots burned in July 1981 – solid symbols; unburned plots – open symbols) at 8 sites.

suggests that two characteristics of fire, in particular, are worthy of further investigation in similar communities. First, how important is seed dispersal from unburnt patches in the re-establishment of a seed bank in the year following fire? In this study by Hassan and West (1986), data were collected from matched pairs of adjacent burned and unburned patches of vegetation left after the wildfire. Second, how would the seed bank be affected by a fire earlier or later in the year? Presumably a fire earlier in the summer would kill adult plants after the depletion of the previous year's seed bank and before the current year's input had started.

This effect of altering the season of burning has been shown for the annual grass, *Sorghum intrans*, in the monsoonal belt of northern

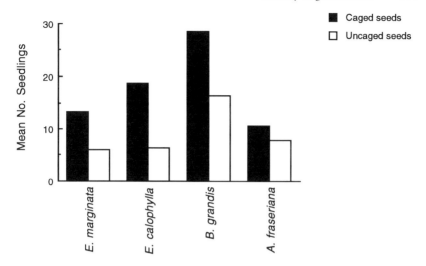

Figure 5.12 Mean numbers of seedlings appearing in caged (solid bars) and uncaged (open bars) treatments in Western Australian eucalypt forest (data from Abbott 1984b). Thirty seeds were placed into caged and uncaged plots in autumn (April) and seedlings counted in summer (January), by which time all intact seeds are expected to have germinated.

Australia (Stocker and Sturtz 1966, Stocker 1966, Gill 1981b). Annual or biennial burning in the dry season does not alter the abundance of this species, at least in the short term, indicating that soil-stored seeds are protected from fire (Andrew and Mott 1983). A single fire in the wet season can devastate the *S. intrans* population, because adult plants are killed at a time when there is no seed bank.

The above examples represent the simplest plant life-history; namely an annual plant with an annual seed bank. Populations of sprouter species that lack seed dormancy will be less conspicuously influenced by fires in the long term. Nevertheless, timing of fire can determine the success or failure of a particular cohort of seeds. For example, a fire just prior to seed dispersal may kill seeds before any reach the safety of the soil, as was the case for the annual grasses.

Some studies suggest that seed predators are a substantial source of seed loss from a seed bank for plant species that release non–dormant seeds annually. For several species in Western Australian eucalypt forest, Abbott (1984b) estimated that predation by seed-eaters between dispersal in April (autumn) and the following January (by which time most intact seeds were expected to have germinated) reduced seedling densities by half (Fig 5.12). This study was conducted in an unburned site. What effect would fire have on these rates of seed predation? If a fire affects the abundance or activity of seed-eaters without reducing post-fire seed production, more germination may be expected due to better seed survival. However, Abbott (1985) reported that mature trees of one of the species studied in relation to seed predation, *Banksia grandis*, failed to reproduce for at least 3 years after a moderate-intensity fire (see Table 5.1). Thus, any advantage flowing from an immediate post-fire decline of seed predators would have been lost. Fire intensity is clearly important here, because low-intensity fires do not seem to affect flowering in *Banksia grandis* and other *Banksia* species that occur in the same areas (pers. obs.). In the light of the potential importance of seed-eaters in dispensing with dispersed seed in unburned vegetation, further studies comparing their impacts in burned and unburned vegetation are warranted.

A warning!

Just because seeds of a species germinate readily in laboratory trials following collection in the field does not necessarily mean they do not display dormancy in nature. Several studies in which germination has been assessed in field-collected soil samples have revealed species that were thought not to display dormancy and should therefore have been absent from the seed bank (e.g. *Salix bebbiana*; Fyles 1989). Moreover, canopy-stored seeds exhibiting no dormancy when they are collected and tested under standard conditions may nevertheless be trapped in cones or fruits for a long time, even for many years (R. Muston pers. commun.), after fires caused rupture of scales or follicles. Enforced dormancy may provide a soil- or plant-stored seed bank of seeds that have no *innate* dormancy.

Species with a soil-stored seed bank

Plant species that produce seeds lacking any innate dormancy (see Harper 1977 for explanation of dormancy terms) may nevertheless develop a seed bank if seeds remain alive but unable to germinate because of inappropriate conditions. Burial in the soil, for example, may enforce dormancy in seeds of such species. The presence of an extremely large seed bank of herbaceous and weedy species stored in the soil is a common observation in many plant communities (Thompson 1978), including those in most regions subjected to periodic fires (Carrol and Ashton 1965, Howard 1973, Vlahos and Bell 1986).

In addition to this dormancy enforced by the physical and/or chemical microenvironment of the seed, many species in fire-prone ecosystems produce seeds with an innate dormancy, usually imposed by an impermeable seed coat. For these seeds, germination follows the application of an appropriate stimulus, commonly fire but, in inter-fire periods, perhaps also scarification or unsuccessful insect attack.

As seeds are made up of living tissue, they are susceptible to lethal temperatures reached in a fire. Thus, fire intensity affects survival of seeds. As seen in Chapters 2 and 3, the temperature actually reached in a fire varies with height above the ground and depth below ground, so the extent of mortality of a seed bank is affected by the distribution of seeds in a height- or depth-profile. Many of the seeds lying in the litter are likely to be consumed in a fire. However, buried seeds and seeds protected in closed fruits in plant canopies are more likely to tolerate fire.

Auld (1986a,b, 1987) conducted a detailed study on the effects of fire on the seed bank of a leguminous shrub, *Acacia suaveolens*, in eastern Australia. Seeds of this species possess elaiosomes that attract ants. Different species of ants handle seeds differently (Hughes and Westoby 1992). Some (typically the genus *Pheidole*) collect seeds, store them in cachés and eat both elaiosome and seed. Some seeds escape, as is typical with many scatter-hoarding mammal and bird species. Others (typically the genera *Aphaenogaster* and *Rhytidoponera*) discard the seed, either on the surface or in the nest, after removing the elaiosome. Auld estimated that 96% of seeds were dispersed (see also Hughes and Westoby 1990). Most of these were encountered by ants and 38% of the total seed fall was estimated to have been collected by one genus of ant (*Pheidole*). With other agents of burial, about 35% of seeds were estimated to have been buried at a depth of 5 cm or greater. In the absence of fire, the decay rate of seeds in the soil was low, with a half-life estimated at over 10 years. Fire had two effects on the seed banks. Heating to 100 °C for moderate periods (i.e. > 60 min.) or to higher temperatures for shorter times caused losses to the seed bank by killing a high proportion of seeds. Less-severe heat exposure caused losses to the seed bank by stimulating germination.

From these measurements, it is clear that successful germination is related to depth of the seed in the soil profile and to the intensity of the fire (Fig. 5.13). From simulated hot fires produced using a gas burner, Auld estimated that maximum germination would be achieved at depth of 3–4 cm (81% of seeds). Below this depth, heating was insufficient to stimulate germination and above it, surface temperatures were sufficient to kill a high proportion of seeds. In simulated cool fires, the maximum

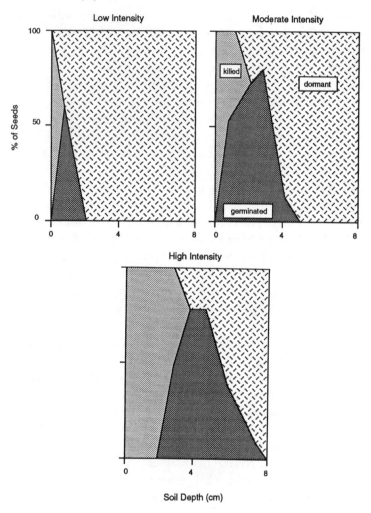

Figure 5.13 Relationship between fates of *Acacia suaveolens* seeds (dark shading – stimulated to germinate; stippling – killed by heat; cross-hatching – remaining viable but dormant) and fire intensity (from Auld 1987 – cf. general relationship presented in Fig. 3.11).

germination rate (60% of seeds) would be achieved from seeds in the top 1 cm. In either sort of fire, the 35% of seeds buried below 5 cm were effectively lost from the seed bank, since even hot fires failed to produce germination from this depth or lower.

The effects of variation in fire intensity are well demonstrated in studies of population dynamics of the introduced grass, *Bromus tectorum*, in sagebrush (*Artemisia* species) grasslands of western North America (Young and Evans 1978, 1985). *Artemisia* shrubs create the spatial structure of these grasslands, because the accumulation of litter and nutrients and the more favourable microenvironment under the shrubs supports greater growth of herbaceous species there than in the inter-shrub spaces. Consequently, the majority of the *Bromus* seed bank is located under shrubs. Wildfires burn the shrub clumps and associated herbs and litter with sufficient intensity and duration to destroy *Bromus* seeds. However, seeds surviving in the inter-shrub spaces produce a plant density of about 10 per square metre in the year after fire (Young and Evans 1978). Massive seed production by these plants produces a density of 2500 to 3500 per square metre in the second year after fire, concentrated in the haloes of ash surrounding the burnt *Artemisia* shrubs. The small-scale heterogeneity of fire intensity influences micro-topographic patterns of soil nutrient status in chaparral (Rice 1993), and this appears to affect the spatial pattern of post-fire regeneration. Using experimental additions of fuel, to increase the intensity of winter fires in localized patches, Moreno and Oechel (1991) found that germination of various herb and shrub species was differentially affected by altering intensity.

Species with a plant-stored seed bank
Seeds held in the canopy of a plant are likely to be exposed to a short burst of high temperature as the fire front passes. The actual temperature achieved will depend primarily on fire intensity and on height of the seeds above the ground. In many species of plants exhibiting delayed seed release (bradyspory), seeds are actually held in capsules, follicles or other woody structures that provide protection from radiant heat. A good example is *Banksia ericifolia* in eastern Australia, which is an obligate-seeding, bradysporous shrub species. Exposed seeds of this species die if a temperature of 150 °C or more is applied for 7 min. (Siddiqi et al. 1976). Bradstock and Myerscough (1981) found that the peak temperature reached in fires in *B. ericifolia* stands frequently exceeded 592 °C, the upper limit detectable in their tests, and other studies indicate that typical peak temperatures may approach 800–1000 °C. Because seedlings appear in *B. ericifolia* stands after fire, we can infer that the peak temperature achieved in the flames is not transmitted directly to the seeds. However, attempts to measure the temperature actually suffered by seeds enclosed in woody follicles under a range of applied fire intensities have been rare. What proportion of a canopy-

stored seed bank dies as a result of excessive heat? This question would not be particularly difficult to study, yet few data are currently available.

One variable that could have a marked effect on the proportion of a *Banksia* seed bank surviving a fire is the age of infructescences. Fruits that are 1 year old or less still retain substantial moisture, while older fruits have dried out. This results in easily observed differences between old and new fruits in the proportion opening after fire and in the time taken until all seeds are released (Cowling and Lamont 1984). Although these differences are most likely to result from different rates of drying of follicles after fire, different moisture levels in old and young fruits may also influence heat transmission to the enclosed seeds. Some infruct escences burn completely in a fire, and older infructescences may suffer disproportionately because of their dryness.

This line of research is well worth further investigation, because seed input to a burned site is likely to be affected by an interaction between fire intensity and residence time of peak temperature, the height-distribution of seeds in the canopy and the proportions of the seed bank represented in confructescences of different ages.

Losses of seeds from a seed bank accumulating after the previous fire may occur through various agencies. Quantitative studies of the dynamics of plant-stored seed banks, such as those conducted by Auld (1987) for soil-stored *Acacia suaveolens* seeds, are badly needed here. They would be relatively simple to carry out, because of the accessibility of plant-stored seeds. Seed viability may decline simply through ageing, but pre-disperal consumption of seeds by both insects and vertebrates may also be of major importance. Whelan (unpublished data), for example, found that only 4% of seeds in 1-year-old confructescences of *B. paludosa* had been eaten by lepidopteran larvae, at the time of a summer wildfire, compared with over 60% of seeds in a sample of older confructescences. It is not known whether this difference in predation between two age-classes of seeds represents a progressive decline in a cohort of seeds or simply different attack rates by the insect in different years. However, a progressive loss due to seed-eating insects is exacerbated by destruction of confructescences by parrots, which eat either seeds or insect larvae (Scott 1982, Scott and Black 1981). Support for these findings comes from studies on South African Proteaceae (Bond 1985). For one species of *Protea* and two of *Leucadendron*, seed viability was significantly lower in 2-year-old than in 1-year-old cones. For five of the ten species surveyed, 2-year-old and older cones contributed less than 50% to the stored seed reserves.

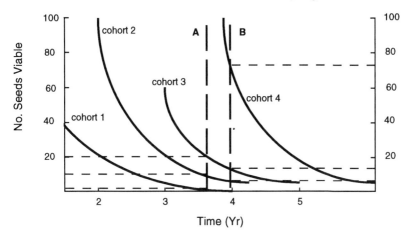

Figure 5.14 Model of seed bank dynamics of a canopy-stored seed bank, assuming a rapid decline in seed viability of stored seeds. Marked annual variations in flowering and seed production are expected. A fire before seed maturation at the end of the third year (A) would release seeds from cohorts 1, 2, and 3 (2, 10 and 20 seeds respectively) and kill the current year's undeveloped crop. In contrast, a fire just after seed maturation (B) would release seeds from cohorts 2, 3 and 4 (5, 15 and 75 seeds respectively).

If the loss of seeds from a cohort stored in a seed bank is progressive (especially if it follows a form something like a negative exponential function), then the timing of a fire in relation to fruit maturation may be a major influence on seedling production by the population (see Jordaan 1949, 1965, Midgley 1989). A fire occurring just after fruits have matured will release, intact, the cohort of seeds that can contribute most to post-fire recruitment. A fire just prior to fruit maturation will exclude the current year's reproduction and release cohorts of seeds that have been subjected to one or more years of progressive predation. Fig. 5.14 illustrates this. The negative exponential lines represent the declining numbers of viable seeds remaining in each annual cohort of seeds. A fire at time A would release a total of 32 seeds – some from each of the three cohorts (indicated by the dashed horizontal lines which mark the interception of the seed decay curves and the time of fire). A fire at time B, in contrast, would release 95 seeds from three cohorts. The conclusions drawn from this sort of model will obviously vary depending upon the shapes of the seed decay curves and the magnitude of each year's seed production. Empirical data are needed here and are not too difficult to collect.

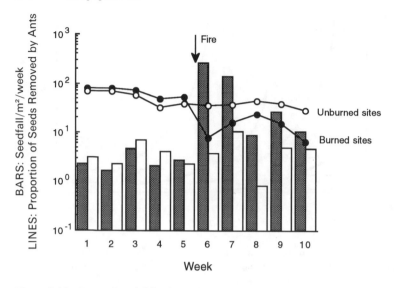

Figure 5.15 Rate of seed fall of *Eucalyptus delegatensis* in burned (shaded bars) and unburned (open bars) sites, and percentage of seed removed from cachés by ants in burned (solid symbols) and unburned (open symbols) sites. Data are from 5 weeks before and 5 weeks after an experimental fire occurring between weeks 5 and 6 (from O'Dowd and Gill 1984). Note the logarithmic Y-axis.

The seed-bank dynamics of bradysporous species are not confined to the time seeds spend in the canopy. Seeds released *en masse* after fire are no longer protected by cone scales or follicles. Although these seeds are now exposed to the dangers described for species with annual seed release and no dormancy (see above), there are important differences. First, the populations of seed-eating animals are likely to differ from those present in the pre-fire vegetation (van Hensbergen *et al.* 1992; see also Chapter 6). Second, synchronized release of abundant seed is equivalent to mast fruiting. Detailed studies of the consequences of this phenomenon for seed survival have been conducted on removal of seeds by ants in eucalypt forests (Ashton 1979, O'Dowd and Gill 1984, Wellington and Noble 1985b). These studies have shown that the lack of appearance of eucalypt seedlings in the absence of fire is correlated with high rates of removal of seeds by ants. O'Dowd and Gill (1984) showed that a high-intensity fire in *Eucalyptus delegatensis* forest stimulated massive seed release and increased ant activity on the soil surface. However, the increase in ant activity was insufficient to cope with the large additions to the seed resource, resulting in lower rates of seed removal (Fig. 5.15). This would lead to a greater probability that some seeds will either be incorporated into the soil profile or germinate before they are found by ants. Wellington and Noble (1985b) substantiated this finding with an elegant experiment in which seeds were added to unburned sites in a simulation of massive seed release following a fire. This addition of seeds to the surface produced a 'pulse' of seeds appearing in the soil profile lasting for some 150 days and producing substantial germination. Seed storage and germination in the control sites were negligible.

Predation of eucalypt seeds by ants may be a special case, because ant populations are so tolerant of fire. However, *Watsonia borbonica* in South African fynbos flowers strongly in response to fire. Le Maitre and Brown (1992), found that the high density of flowering did not satiate pollinators, but did result in a substantial reduction of pre- and post-dispersal seed predation by weevils (Curculionidae). As an aside, Brewer and Platt (1994) also refer to an example of satiation of herbivores. White-tailed deer in Florida were apparently satiated by the massive production of shoots by the forb *Pityopsis graminifolia* after spring fires (in May). Fires in other seasons did not have the same effect. Very little information is available concerning post-fire consumption of seeds by vertebrates such as birds and mammals. Perhaps fires that fail to reduce populations of seed-eating vertebrates permit a substantial decline in seed numbers as seed-eaters concentrate on the the abundant resource released in the burned patches. This suggestion is supported by South African studies of rodent predation on seeds of *Protea repens*, which are released by fire (Bond 1984; but see Midgley and Clayton 1990 for contrasting results). Over 15 weeks after a spring fire, about 75% of seeds buried in trays were harvested by rodents. Only 9% of seeds in caged trays were removed, none apparently harvested by rodents. This finding introduces another likely effect of season of burning on the seed bank. Because seeds are so susceptible to rodent predation, the length of time they spend on the ground after being released from the relative protection of storage on the plant and before escaping predation by germinating is probably crucial in determining subsequent seedling densities. Most of the Proteaceae species in the Cape fynbos delay germination after seed release until autumn. Thus, seeds released after autumn fires will be able to germinate immediately. Seeds released after spring fires will be exposed to rodents for up to 25 weeks prior to germination.

Seed dispersal

Seed dispersal in relation to fire is a topic that was reviewed by Whelan (1986). Seed dispersal from outside a burned site appears to be of over-riding importance in colonization for only some species in particular environments (see Chapter 3). For example, Rowe and Scotter (1973) listed several species in high latitudes that can only reappear in a site following seed invasions from unburned plants at the boundaries of the fire. In floras with a high proportion of such species, the pattern of community change after fire may be determined predominantly by differential invasion rates.

For many fire-prone plant communities, most seedlings appear to originate from seeds already present within the burned site (Whelan 1986). Seed dispersal in these situations is important in two contexts. First, annual dispersal in the absence of fire can determine the spatial dispersion patterns present at the time of fire: e.g. the depth of burial, distance of seeds from parent plants and other vegetation etc. Second, post-fire dispersal of seeds, both from canopy-borne fruits or cones and from the first flowering of sprouting adults or new seedlings, can be enhanced by the more open vegetation structure, with higher wind speeds and less obstruction. The consequences of these patterns of seed dispersal for plant population dynamics have not really been addressed.

Seedling survivorship and establishment

The Type III survivorship curve (Deevey 1947) represents a probability of death that is very high among young plants and declines with age. This is typical of survivorship curves for single cohorts of many plant populations, especially of trees, followed over time. It is also typical of static age distributions for a variety of species (see Harper 1977, Silvertown 1982). Factors that affect establishment and mortality among young plants are important because the enormous numbers of seedlings present per adult plant would permit a rapid, massive increase in population size if they were all to survive. In obligate-seeder species, the whole post-fire population must come through the seedling stage of the life-history because wildfire causes mortality of established plants and germination of new seedlings appears to be confined to the immediate post-fire period. Thus, factors determining seedling mortality will be important not only in explaining post-fire population sizes but also in determining population persistence versus local extinction. For sprouting species, in contrast, maintenance of the established plants through fire provides a buffer, or lower limit, to population decline due to poor recruitment. Population increases will be regulated by survivorship patterns of post-fire cohorts of seedlings.

A variety of factors can affect the pattern of seedling survivorship of a cohort. The shape of the survivorship curve, with seedlings at high density displaying a high rate of mortality, declining as plants age and density falls, suggests that density-dependent factors may be important in determining population fluctuations. Density-dependent factors could include inter- and intra-specific competition, herbivory and pathogen infections. The relatively high mortality over summer periods found in many studies indicates the importance of desiccation and other factors related to a long dry season. These mortality factors may operate in a density-independent fashion. It is important to understand that the type-III survivorship curve is not proof of density-dependent processes in action. The risk of mortality from a density-independent factor, such as drought or trampling, may depend upon a plant's size or life-history stage. Mortality rate may therefore be correlated with density only because plants are becoming more tolerant of density-independent factors as they age.

In order to understand the effects of a fire on recruitment to populations, we need to know whether survivorship patterns differ between the burned site and an unburned site, or how survivorship patterns vary over time since a fire (e.g. Gray 1987; see Fig. 4.4). However, for some plant species few if any seeds germinate in unburned sites, so this comparison is not possible. In any case, investigating the effects of fire regime will require comparisons of survivorship after fires of different intensity, season or extent. The published literature includes very few examples of this approach. The following section examines survivorship of seedlings in burned and unburned sites and the various factors that may determine recruitment in post-fire cohorts of seedlings.

Fire and survivorship

The post-fire environment appears to be favourable for seedling survival and growth if seeds can survive the passage of fire or get to the site soon afterwards. Whelan and Main (1979), for example, showed that the survivorship curves for seedlings germinating in south-west Australian eucalypt woodland were considerably steeper for unburned sites than for burned sites (Fig. 5.16). Furthermore, there was a distinct seasonal component to mortality in this Mediterranean-climate region, with most mortality occurring over the summer drought period. It is also interesting to note that the apparently favourable post-fire conditions had disappeared by the second year. Survivorship in burned and

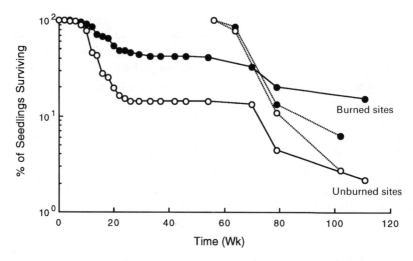

Figure 5.16 Survivorship curves for two years of seedling production (all species lumped) in Western Australian eucalypt woodland in burned (solid symbols) and unburned (open symbols) sites after experimental fires in March 1975 (data from Whelan and Main 1979). Initial seedling densities were an order of magnitude higher in the burned sites than in the unburned sites in the year of the fires (1974 versus 91 seedlings), due to stimulation of germination of several species of shrubs, but less different in the following year (899 versus 357).

unburned sites was similar by this time. Purdie (1977) reported similar results in eastern Australia (Fig. 5.17). Seedling mortalities for most groups of plants were higher in unburned than in burned sites. Moreover, in both burned and unburned sites, seedlings reaching the second year ran a lower risk of mortality than during the first year after germination.

Of course, the season of burning might be expected to have an impact on the survivorship of seedlings that appear afterwards. Brewer and Platt (in press) found in Florida that seedling survivorship was greater after experimental May fires (spring) than after experimental fires in other months.

Density-dependent limits to survival

There are few studies of density-dependent thinning in plant populations in the field (Antonovics and Levin 1980). However, the fact that massive seed release or a large pulse of germination may follow fire, as discussed in the previous sections, indicates that seedling densities may indeed

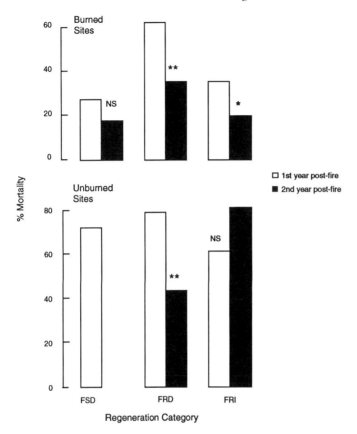

Figure 5.17 Annual mortalities in a cohort of seedlings during the first (open bars) and second (solid bars) year after fire in burned and unburned sites (data from Purdie 1977). FSD – fire-sensitive decreasers; FRD – fire-resistant decreasers; FRI – fire-resistant increasers. Asterisks by the bars represent the level of significance of the differences: NS $P > 0.05$; $\star$ $0.05 > P > 0.01$; $\star\star$ $0.01 > P > 0.001$.

become so high that competition and other density-dependent processes inevitably ensue.

There are some indications that this may be confined to particular circumstances. Seed and therefore seedling densities are spatially highly variable in nature. Seed dispersal typically produces high densities of seeds near parent plants. Secondary dispersal by wind or surface runoff may produce local concentrations of seeds in topographic depressions or against obstructions (Whelan 1986, Lamont *et al.* 1993). As discussed in the previous section, seed cachés made by ants or small mammals may

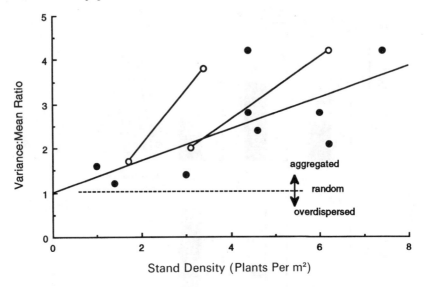

Figure 5.18 As stands of *Ceanothus megacarpus* age, so density declines (from Schlesinger and Gill 1978). The highly aggregated distribution of seedlings at high density becomes more random as the stand thins (regression line). The two pairs of connected, open symbols represent the only two stands for which patchiness and density were measured at two dates: other data points are for separate stands of differing post-fire age. The horizontal dashed line represents the variance:mean ratio for a random spatial pattern.

result in dense clumps of seedlings far removed from parent plants. How frequently are seedling densities high enough to result in density-dependent mortality?

One approach to this question has been to investigate changes in spatial pattern of a cohort of seedlings over time after fire. If density-dependent mortality were operating, individual seedlings in dense clumps would sustain a higher risk of mortality than isolated seedlings, producing an increasing evenness in spatial pattern over time. It is this process that has been used to explain the regular spacing patterns that have been observed for desert shrubs (King and Woodell 1973, Antonovics and Levin 1980). Schlesinger and Gill (1978) showed that the variance:mean ratio for the density of *Ceanothus megacarpus* plants in 80, 1 m² quadrats approached unity (= random spatial pattern) in the oldest, least-dense stands. For the two stands for which current density and distribution could be compared with the original post-fire densities, spatial pattern changed from highly aggregated after fire to nearly random as the density declined over time (Fig. 5.18). Similar results were obtained in a gorse shrubland in Spain (Carriera *et al.* 1992). These findings imply that density-dependent mortality was occurring.

Whelan (1986) also used this inferential technique to explore density-dependent mortality among seedlings regenerating after fire in Western Australian eucalypt woodlands. Seedlings were tagged and mortality monitored over 25 months in 200, 1 m^2 quadrats per plot. Patchiness was expressed by Lloyd's Index (Lloyd 1967, Pielou 1974). This index measures pattern independently of seedling density. *Random* disappearance of individual seedlings would alter density without affecting patchiness. In contrast to the expected pattern of decreasing patchiness over time, seedlings of the three shrub species examined (Fig. 5.19) showed a general increase in patchiness as density declined. Close inspection of these data reveals some interesting patterns. During the periods of most severe mortality, namely the two summers (months 2–6 and 14–18 in Fig. 5.19), patchiness actually increased markedly for all three species in each plot. Where mortality did occur over winter and spring, however, as with *Jacksonia sternbergiana*, patchiness did decline as expected. These results suggest that different mortality processes may be operating at different times of year. In summer, the seedlings that die are generally isolated individuals rather than those in clumps. Perhaps the microsites that are favourable for germination also favour survival over the summer. However, in the cooler, growing months, density-dependent mortality did seem to be occurring.

Because of the insensitivity of this technique, it is still possible that some particularly dense clumps of seedlings exhibit density-dependent mortality that is not detected in an index of aggregation. More direct approaches to the question are needed. A good model is a study by Wellington and Noble (1985a), who examined density-dependent mortality directly in cohorts of eucalypt seedlings appearing after fire in a mallee community in semi-arid Australia. Survivorship curves were calculated for samples of seedlings occurring naturally in three density-classes, > 20, 5–20 and < 5 seedlings per square metre. General trends in the curves were similar, with most mortality occurring over the summer period, but greatest mortality rates were clearly recorded for the highest density-class and lowest rates among the sparse seedlings (Fig. 5.20). This finding is supported by a study by Lamont *et al.* (1993), who experimentally thinned seedlings growing in dense stands in litter-rich microsites. They found that percentage survival of seedlings over the first year after germination was much higher for seedlings growing in low densities outside these clumps than for seedlings growing in clumps. Thinning of

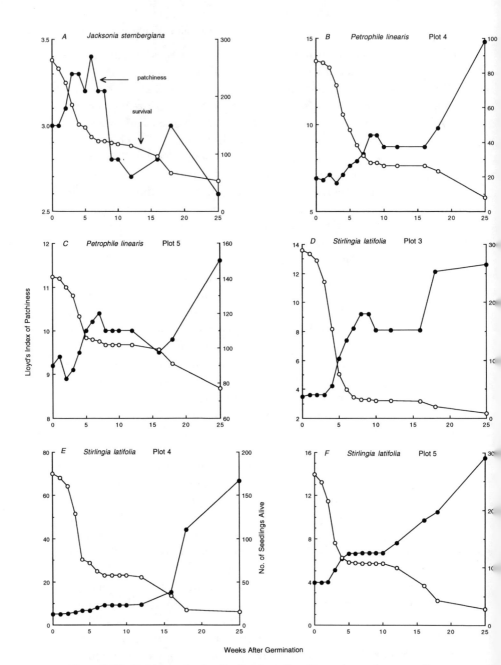

Figure 5.19 Patchiness in the distribution of seedlings changed as mortality occurred over time in south-west Australian eucalypt woodland (data from Whelan 1977). Patchiness, measured by Lloyd's Index (Pielou 1974, p. 152) generally increased (cf. Fig. 5.18).

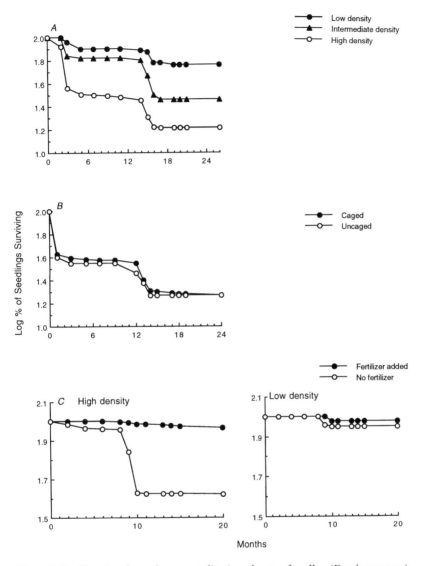

Figure 5.20 Density-dependent mortality in cohorts of mallee (*Eucalyptus* spp.) seedlings germinating after fire in western New South Wales, Australia (data from Wellington and Noble 1985b).
A. Survivorship was greatest for seedlings at low density and least for those at high density.
B. Protection from herbivory by cages did not influence survivorship.
C. Experimental addition of fertilizer increased survival at high density (left) but not at low density (right).

seedling densities within the clumps substantially improved survivorship.

Even if density-dependent mortality is implicated in regulating
population growth, as in some of the above examples, there can be
various causes of the mortality; intra-specific competition, inter-specific
competition, herbivory and pathogens.

Intra-specific competition

Understanding the mechanisms of intra-specific competition has been
central to the development of agronomy and plant population biology
(Harper 1977), and we easily develop an intuitive feeling that competition must be important in natural systems. However, the evidence that
mortality within cohorts of seedlings is primarily a result of intra-specific
competition is rather scanty, especially from field studies.

There has been a great deal of interest and debate about the nature of
self thinning in plant populations, focussed on the degree to which the
$-3/2$ power law is a good description of the relationship between mean
plant weight and density. As plants in a population grow in size, their
density declines due to mortality. However, mean plant weight increases
at a faster rate than the decline in density, producing the slope near to
$-3/2$. The main features of this process are as follows (Silvertown 1982):

(i) Plants increase in mean size (weight) over time;
(ii) No density-dependent mortality occurs until populations reach the
self-thinning line;
(iii) Mortality begins sooner in dense populations than in sparse ones;
(iv) Plants of the same weight will be younger in a sparse population
than in a dense one;
(v) All populations will eventually reach a stage where weight
increment and mortality are balanced (slope $= -1$ rather than $-2/3$). The increase in total plant weight in the population ceases at this
point;
(vi) Dense populations reach a slope of -1 before sparse ones.

This principle is illustrated in Fig. 5.21, which relates tree density to
tree volume in ten stands of ponderosa pine, at different stages of post-
fire regeneration. Schlesinger and Gill (1978) found a slope of -1.23 for
10 stands of *Ceanothus megacarpus* in California chaparral. Like ponderosa
pine, this obligate-seeder shrub forms even-aged stands of seedlings after
fire kills the parent plants and releases the seed bank from dormancy.

A danger inherent in this inferential approach to detecting intra-

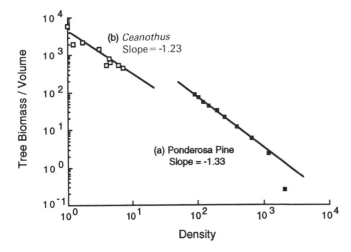

Figure 5.21 Self-thinning lines that approximate the expected slope of −1.5 found for fire-initiated stands of (a) ponderosa pine (White 1980) and (b) *Ceanothus megacarpus* (Schlesinger and Gill 1978). Ponderosa pine data relate density in trees acre⁻¹ to mean tree volume in ft³. *Ceanothus* data relate density in shrubs m⁻² to mean shrub biomass in g dry-weight.

specific competition should be pointed out. Gill and McMahon (1986) discovered a −3/2 power relationship between estimates of plant size and measurements of density of *Banksia ornata* populations of different post-fire ages and they used this adherence to the −3/2 power law as evidence that density-dependent thinning explained population changes over time since fire. However, the intercept falls far outside the expected range for any species (see White 1980). It is not clear why this might be so, but as emphasized earlier in this chapter, a chronosequence of sites might be a poor indicator of temporal change within a single site.

Morris and Myerscough (1988) have conducted one of the few diachronic, experimental studies of self-thinning in a natural plant population recovering after fire. Their system was the obligate-seeder, *Banksia ericifolia*, in eastern Australian eucalypt forest and heath. This makes a good parallel with the Schlesinger and Gill's study on *Ceanothus*, described above. High densities of *Banksia* seedlings resulting from fire-stimulated seed release were manipulated to produce two densities in two classes of plots: high growth and low growth. In all but the high-density, high-growth plots, survivorship curves approximated straight lines, implying a nearly constant rate of mortality over time, regardless of density (Fig. 5.22*A*). There was, however, some seasonal variation this

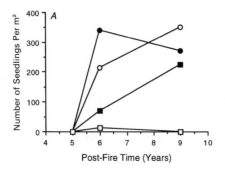

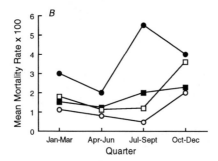

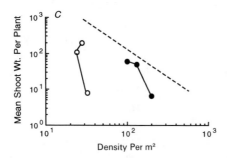

Figure 5.22 *A*. Survivorship curves for samples of *Banksia ericifolia* seedlings
manipulated to produce high (solid symbols) and low densities (open symbols) at
sites of high productivity (circles) and low productivity (squares) (data from
Morris and Myerscough 1988).
B. The spring–summer quarter produced the highest rate of mortality in all
treatments except high–density, high–growth (symbols as for *A*).
C. Log weight:log density plot for high–density, high–growth plots (solid
symbols) and high–density, low–growth plots (open symbols). The dashed line
represents the − 3/2 thinning line with the intercept proposed by Morris and
Myerscough (1988).

density-independent mortality, with the highest risk of death occurring in the October–December quarter (spring–summer) for all but the high-density, high-growth plots (Fig. 5.22B). The high-density, high-growth plots thinned at a faster rate than the other treatments (Fig. 5.22A) and the log weight:log density relationship approximated the expectation of the $-3/2$ power law (Fig. 5.22C), implying that density-dependent intra-specific competition had begun to occur. It is interesting to note that this cause of mortality was concentrated in the winter–spring (July–Sept.) quarter of the year (Fig. 5.22B). Studies such as this may indeed reveal that intra-specific competition is important in population regulation. However, Morris and Myerscough's study does indicate that this factor is not important throughout a regenerating stand of B. ericifolia, at least during the 6 years of study after fire; rather it occurred within the relatively few dense patches of seedlings growing in favourable micro-sites. Plant deaths caused by factors such as herbivory, trampling, climatic extremes or inter-specific competition must be responsible for the density-independent mortality observed in other plots.

Studies that focus on obligate-seeder species, such as ponderosa pine, *Banksia* and *Ceanothus* described above, had to deal only with a population consisting of a single cohort of plants. This is unlikely to typify many species in fire-prone environments. The post-fire population will usually be a mixture of surviving adult plants and new seedlings, also interspersed among representatives of any number of other species. Because it will be difficult to incorporate these variables into a basic self-thinning model, the more direct approach of field experiments on naturally occurring seedlings may be of greater value.

The study by Wellington and Noble (1985b) provides direct evidence for the importance of competition, especially intra-specific competition, in post-fire seedling mortality. Experimental additions of water and fertilizers to seedlings growing at high density (>20 per m^2) significantly increased survivorship over high-density control groups while water and fertilizer additions failed to affect survivorship of low-density seedlings (<5 per m^2) (see Fig. 5.20).

The general conclusion of the studies outlined above is that intra-specific competition can indeed control population size under certain circumstances. Where massive germination after fire produces dense, monospecific stands of seedlings that are relatively unaffected by other causes of mortality, self-thinning is likely to occur at some stage of post-fire regeneration. Population density will then be regulated by the carrying capacity of the site for mature individuals. However, most post-fire vegetation is a mixture of species and, within each species, a mixture of sprouts and seedling cohorts, thus making it difficult to detect the ultimate causes of mortality. Moreover, other mortality factors, such as herbivory, pathogens and climatic extremes may thin a stand of seedlings well before they interfere with each other or with individuals of other species.

Inter-specific competition

How does inter-specific competition affect post-fire population dynamics. This question is of particular relevance in situations where a post-fire succession occurs and the 'pioneer' species may either enhance or inhibit the occupation of the site by later successional stages. Nevertheless, there appears to have been little detailed work on population dynamics in relation to inter-specific competition. Inter-specific competition has been invoked as a process that explains the relative importance of obligate-seeder and sprouter species of shrubs in several shrublands of the world (Parkes 1984, Keeley 1977, Specht 1981, Smith *et al.* 1992). However, studies of root architecture and water relations have generally failed to find adaptations that might explain co-occurrence or dominance of one group over the other (Smith *et al.* 1992).

Effects of herbivores

Herbivores can be a major cause of seedling mortality and they may regulate population size by operating in a density-dependent manner. Whelan and Main (1979) showed that the mortality of seedlings caused by invertebrate herbivores over two summers after fire was less than that in matched unburned sites. Furthermore, most mortality due to this herbivory occurred during the first year after fire (Tables 5.3, 5.4). Those seedlings that survived into the second year were more tolerant of herbivory. Whelan and Main's study also illustrated another important point – the impact of insect herbivores varied among plant species. Legume seedlings, in general, appeared to be more susceptible to death from insect herbivory than the other common species that were in the family Proteaceae.

Although these observations provide good evidence that seed- and seedling-eaters do kill seedlings, and indicate that herbivory is more intense in unburned vegetation, one should guard against using this as evidence for herbivory as a consistent *regulator* of plant population size. Although 11% of seedlings in unburned plots and 3% in burned plots in Whelan and Main's study were killed by insect herbivores over the first

Table 5.3. *Mortality due to herbivory by insects over one year in the post-fire cohorts of seedlings of nine shrub and tree species in south-west Australian eucalypt woodland*

Family/species	Numbers of dead seedlings		% eaten
	Not eaten	Eaten	
Proteaceae			
Banksia attenuata	16	0	0
B. menziesii	10	1	9.1
Petrophile linearis	360	9	2.4
Stirlingia latifolia	920	7	0.8
Mimosaceae			
Acacia stenoptera	165	3	1.8
Fabaceae			
Hardenbergia comptoniana	35	4	10.3
Jacksonia sternbergiana	262	29	10.0
Kennedia prostrata	19	0	0
Myrtaceae			
Eucalyptus marginata	6	1	14.3

Source: Data from Whelan and Main (1979).

Table 5.4. *Mortality, over two consecutive years, in a cohort of* Jacksonia sternbergiana *seedlings that germinated after a fire in 1975*

	Year 1	Year 2
Number alive at start of year	292	159
Number dying during year	133	53
Number of these deaths due to herbivory	32	4
% of deaths caused by herbivory	24.1	7.5

Source: Data from Whelan and Main (1979).

year, recensusing of the cohort 8 years later revealed that only 3 seedlings out of the initial 2065 were still alive (unpublished data). For most species, including some that sustained almost no herbivore damage, mortality was 100% before any individual reached maturity. This anecdote points to the importance of density–independent, stochastic events in negating the impact of interactions that may be easily measured over a short time span. Herbivores may be eating seedlings that are destined to die of other causes before they reach a reproductive stage.

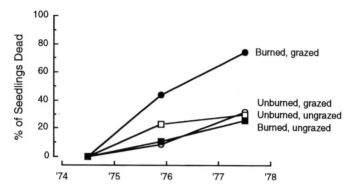

Figure 5.23 The interaction of fire and herbivory produces the greatest seedling mortality of the legume *Daviesia mimosoides* in eastern Australia (data from Leigh and Holgate 1989; cf. data for adult plants Fig. 5.6).

Notwithstanding this warning about the observational approach, the important role of herbivores is illustrated in other studies. Leigh and Holgate (1979) found an interaction between herbivory by mammals and fire for the legume *Daviesia mimosoides* over four years. In 1973, seedlings were tagged in burned and unburned sites, both inside and outside vertebrate-proof exclosures. By 1977, proportions of seedlings surviving were equivalent inside and outside exclosures in unburned sites (where germination was low). In contrast, after fire (where seedling densities were high due to stimulated germination), grazing markedly increased seedling mortality (Fig. 5.23).

Studies on seedling survival of South African Proteaceae contrast with the above findings. Bond (1984) found that although seeds released after fire were highly likely to be consumed by rodents, germination removed them from the risk of herbivory. Equivalent proportions of caged and uncaged seedlings died over a 22-week period. This was not the case in unburned vegetation, however. In 20-year-old heath, nearly 2/3 of seedlings not protected by exclosure cages were eaten by herbivores. Seedlings in exclosures all survived.

In Californian chaparral, where germination of many shrub species appears confined to the post-fire year, herbivory by mammals can reduce seedling survivorship over several years after fire. Mills (1986), for example, showed that over 2 years after fire, 64% of *Ceanothus greggii* seedlings survived in caged, insecticide-treated plots while only 6% survived without protection. *Adenostoma fasciculatum* showed a similar response, with 34% of seedlings surviving inside cages and only 6%

Table 5.5. *Fates of browsed seedlings in the post-fire cohort of two shrub species in Californian chaparral*

	Adenostoma fascicularis		Ceanothus greggii	
	No. browsed by mammals	% dying	No. browsed by mammals	% dying
1982	32	100	54	100
1983	8	12.5	22	50
1984	0	—	60	25

Source: Data from Mills (1986).

without protection. Observations of mortality over three post-fire seasons confirmed the conclusion of Whelan and Main (1979), mentioned above, that seedlings became more resistant to herbivory as they aged (Table 5.5).

Density-independent events: post-fire climate

Post-fire climate is likely to have a strong impact on recruitment to plant populations. Seedlings are particularly susceptible to low water availability, and the high mortality rates occurring over the dry summer period in Mediterranean-type climates. In fact, some characteristics of plants in these ecosystems, such as dispersal and germination cued by fire, have been seen as adaptations that ensure that seedlings appear in autumn, when rainfall is likely in the following months. It seems likely that characteristics such as these would have evolved in situations where the timing of climatic conditions that are suitable for seedling establishment and growth is predictable in relation to the timing of fire. In climatic regions where the timing of both fire and also subsequent favourable environmental conditions are variable and unpredictable, recruitment may be more of a lottery. Success in the lottery would depend on the stochastic association of a fire and favourable conditions. This hypothesis makes post-fire climate a factor of over-riding importance in determining recruitment to plant populations (Midgley 1989, Bradstock and Bedward 1992). It may explain common anecdotal observations of an abundance of seedlings of a given species after one fire but a dearth after another in the same site (e.g. *Banksia paludosa* in south-eastern Australia; R. Carolin pers. commun. and pers. obs.). Testing of this hypothesis will require measurements of germination and survivorship after a series of fires. It has therefore not received the attention its importance demands.·

Long-term studies and models

The multitude of patterns of response to fire displayed by different life-history stages within a population must now be integrated into cohesive models, if predictions about fire effects are to be made.

As stated in the introduction to this chapter, population responses to fire should be examined over a sequence of fires, and not just followed for some time after a single fire. The many examples discussed above reveal that characteristics of fires are highly variable, as are other factors such as herbivory and post-fire climate. Are generalizations possible? One set of constraints on the variety of possible population responses to fire will be imposed by characteristics of the plant species.

The models illustrated in Fig. 5.24, in which the arrows represent the occurrences of fire, relate to the general characteristics of plant species discussed by Noble and Slatyer (1977) and Rowe (1983). Models *A*, *B* and *D* describe species for which the given fire intensity is adequate to kill established plants and population regeneration is therefore dependent upon seed germination. In *A* and *B*, germination is progressive, occurring as seeds appear in the burned site either dispersed from outside (*A*) or stored within the site (*B*). In *D*, germination is confined mostly to the immediate post-fire period. Model *C* represents a tolerant species in which established plants mostly survive fires and population fluctuations are due to variation in germination and establishment of seedlings. Model *E* represents a similar situation in which fire promotes a burst of germination followed by rapid seedling mortality but established plants survive fires well.

Figure 5.24 Models of various population responses to fires for plant species of different life-history characteristics. Population size (*y*-axes) includes seedlings but not soil- or plant-stored seeds.

In *A*, *B* and *D*, fires (indicated by arrows) are of sufficient intensity to kill established plants and population recovery is therefore dependent upon seed germination.

In *A* and *B*, germination is progressive, occurring as seeds appear in the burned site either dispersed from outside or produced by post-fire flowering within the site (e.g. stippled curve in *A*) or stored within the site (solid curves). *B* represents more frequent fires than in *A*, and on-site seed storage.

In *C*, established plants mostly survive fires. Population fluctuations are due to variation in germination and establishment of seedlings and to ongoing mortality

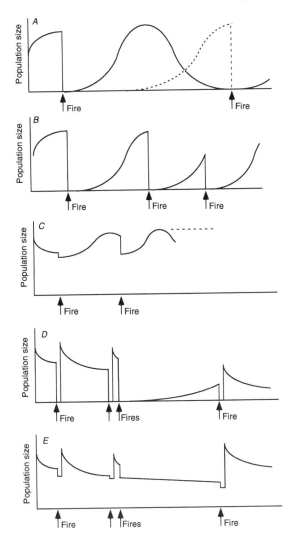

of adults. An upper limit to density (= carrrying capacity) appears as a dashed line.

In *D*, germination is confined mostly to the immediate post-fire period (established plants die as a result of the fire). Two fires in close succession (middle of graph) may remove the obligate seeders and their offspring, resulting in a slow return to the site via dispersal from outside the site.

In *E*, fire promotes a burst of germination followed by rapid seedling mortality (as in *D*) but established plants survive as well. Note that the two fires in close succession here also remove a cohort of recruitment, but do not have the same lasting impact on the population.

The models presented in Fig. 5.24 indicate the likely effects of different fire frequencies, but not other aspects of a fire regime. For example, when population recovery after fire is solely due to seeds dispersed from outside the burned site, population growth after one fire should not be dependent upon the length of time since the previous fire. Rather, it should be related to the proximity of a seed source and the dispersal characteristics of the seeds. If recovery depends on seeds accumulated within the site, post-fire population growth will depend initially upon the time available since the previous fire for build-up of the seed bank. In situations where established plants (and seedlings) die in a fire and regeneration comes from a dormant seed bank released by the fire (model *D*), two fires close together would deplete all representatives of the species (seed, seedling and adult), so producing a slow rate of population growth while invasion took place from outside the site.

Superimposed on any of these models must be other variables, some of them also related to characteristics of fire, such as intensity of herbivory, and others more stochastic, such as the vagaries of post-fire rainfall and temperatures. In most of the models illustrated here, it is clear that population changes in relation to fire display episodes of rapid population growth or decline, rather than gradual, directional changes. Hence, studies of population dynamics in relation to fire will be necessarily complex and must be conducted over long time spans.

Outstanding questions

The following is simply a list of the aspects of plant population responses to fire that appear to be potentially very important, but about which too little is known.

1. How does patchiness of a fire influence the proportion of established plants that survive in a population, especially of obligate-seeder species? Good measurements of the spatial extent of even-agedness in obligate-seeder populations could easily be made.
2. How does patchiness, or extent, of a fire influence the post-fire herbivore–plant interactions?
3. What is the role of pre-fire seed dispersal (giving a particular pattern of dispersion) in survival of the seed bank through a fire? How does post-fire seed dispersal determine seed survival to germination?
4. What are the seed-bank dynamics (in soil, on plant) over a series of fires?

5. Is self-thinning in local sites important in determining post-fire population density? Do causes of seedling mortality vary among seasons? Do herbivores reduce seedling densities enough to preempt self-thinning? Even if herbivores do not reduce the ultimate seedling density below the level set by other factors, do they nevertheless affect the genetic makeup of the cohort?

6. How do sprouter populations respond to a sequence of fires? Are stored reserves and/or a bud-bank depleted by a particular sequence of fires?

7. Do the chance elements of post-fire climate have an over-riding effect on plant population dynamics?

6 · Animal populations

Introduction

The immediate and obvious effect of fire on animal populations is mortality. Inspection of recently burned areas invariably reveals some charred corpses of a wide range of vertebrates and invertebrates (Chew *et al*. 1958, Newsome *et al*. 1975, Main 1981, Recher and Christensen 1981). Wildfires in coastal regions of south-eastern Australia are often followed by the finding of dead birds washed up on beaches. After a wildfire in Tasmanian sclerophyll forest, an ecosystem supporting 72 bird species, corpses of 59 species were found washed up, including strong fliers such as the goshawk, *Accipiter cirrocephalus* (Hemsley 1967). After wildfire in southern New South Wales, individuals of 42 species were found (Recher and Christensen 1981). Chew *et al*. (1958) found 43 mammal carcasses, including rodents, rabbits, an oppossum and a deer, after a fire in Californian chaparral.

These sorts of observations lead to a widely held belief among the general public that wildfire must be wholly intolerable to any animals present. However, Main (1981) argued for a more dispassionate consideration of fire, which should follow from our understanding that this disturbance has been an integral part of many ecosystems for a very long time, and is therefore likely to have stimulated the evolution of fire tolerance (see Chapter 3). Some individual animals do survive the passage of a fire, and it is biologically more sensible, and more dispassionate, to study how animal *populations* respond to different fire regimes.

Kikkawa *et al*. (1979) emphasized that the effects of fire on populations of vertebrates will depend on the various aspects of fire regime: frequency, intensity, extent and season. Unfortunately, few studies have attempted to address animal population dynamics in relation to any of these features of fire regime. Indeed, it will become clear from the following discussion that an application of the theoretical approaches of population dynamics described in Chapter 4 is still some distance in the future of fire ecology. Few generalizations are available. However, it is not the aim of this chapter to provide a wide review of the many descriptions of changes in populations of animals as a result of fire. Instead, this chapter emphasizes and illustrates likely generalizations and highlights profitable avenues for future study.

Mortality caused by fire

Mortality of animals in a fire must instantly reduce population size. However, despite many anecdotal reports as illustrated above, there appear to be very few quantitative data describing the magnitude of the instantaneous population decline during a fire for any animal species. One reason for this, of course, is that most studies of wildfires start *after* the event! As stated by Main (1981), studies in the proximity of a fire front are 'far too dangerous to allocate to that universal workhorse in boring, distasteful or dangerous research – the graduate student'! Simply comparing counts of animals before and after fire, or in nearby burnt and unburnt vegetation, does not distinguish between emigration and mortality. Marking a sample of animals before a wildfire and following their fates afterwards is a valuable approach that has been rarely used, for obvious reasons! The animal groups that are easiest to work with in replicated fire experiments (i.e. animals that are present in high densities), for example invertebrates or small mammals, do not lend themselves to long-term marking and recapture studies because of their small sizes and relatively short lifespans. Despite these difficulties, a number of studies reveal an interesting variety of patterns.

Many studies of mammal populations before and after fire, especially in California, Australia and South Africa (see, *inter alia*, Lawrence 1966, Newsome *et al.* 1975, Recher and Christensen 1981, Bigalke and Willan 1984, Frost 1984, Quinn 1986), relate that mortality in a fire is surprisingly slight and that animals are frequently captured or otherwise recorded soon after fire. In the detailed radio-tracking study on two Western Australian marsupial species described previously (Chapter 3), Christensen (1980) found that only one animal of the 30 being tracked during and after fire died during the fire itself, from suffocation in the hollow log in which it had sought refuge. Lawrence (1966), in a comprehensive study of responses of vertebrates to fire in Californian chaparral, reported that small mammals that were tagged before fire could be recaptured alive only days afterwards.

There are few quantitative investigations of invertebrate mortality in fire, but two comprehensive studies (Y. Gillon 1972, D. Gillon 1972) describe the responses of arthropod populations to a variety of fires in an east African savanna. Y. Gillon reported that adult acridid grasshoppers are conspicuous flying away from the fire front, and virtually no burned animals were found in the cinders behind the front. By comparing the density of grasshoppers remaining alive in the burned area plus the density of burned carcasses with the number of grasshoppers expected in 100 m² of unburned savanna, Gillon inferred that about 88% of the population, both nymphs and adults, were able to flee the fire. Adults were apparently better able to leave the burning area than nymphs, but the numbers of nymphs burned was nevertheless astoundingly small. D. Gillon (1972) studied the responses of a community of pentatomid bugs to a series of fires and came to similar conclusions. A large proportion of individuals fled the fire front, and some apterous (wingless) and winged adults also survived within the area of the fire. It is difficult to know how general these findings might be. The fires in these studies were part of a series of annual, relatively low-intensity burns. A contrast, within Africa, to these findings is found in a study by Gandar (1982) reporting almost complete destruction of the arboreal insect fauna in burning savanna.

Post-fire population changes

One particular difficulty in assessing fire-induced population changes in animal species is the fact that, even in the absence of fire, populations typically undergo marked seasonal and annual fluctuations. Changes after fire must be detected against this background variability. The experimental 'design' of comparing pre-fire and post-fire censuses poses a number of statistical difficulties, discussed in Chapter 4 (see Hurlbert 1984, Stewart-Oaten *et al.* 1986). Few studies of changes in animal populations after fire have even attempted adequate replication or statistical analysis. These shortcomings should be borne in mind when interpreting the results of the various studies reviewed below.

Soil and litter invertebrates

The literature describing the responses of populations of soil and litter invertebrates to fires is characterized by enormous variability. An early review by Ahlgren (1974) strongly suggested that populations of annelids, molluscs, many insect groups, arachnids, and centipedes and millipedes are severely reduced by fire. However, even within each of these groups, there is substantial variability. Surveys of studies of the responses of soil and litter invertebrates to fire (Dolva 1993, Tap 1994, Friend 1994; Table 6.1) reveal that some report an increase in overall invertebrate population in response to fire and others a decrease.

There are many possible reasons for such variability in results. First, the grouping 'soil-and-litter-invertebrates' includes a wide range of organisms, some large and confined to the litter layers, which might burn completely in an intense fire, and others microscopic (e.g. mites and collembola) and found at such a depth in the soil that the fire temperature is hardly felt. Second, information regarding some important aspects of fire regime (intensity, area, patchiness) is rarely recorded and these factors are likely to vary greatly among different studies. Furthermore, studies have been in a very wide variety of ecosystems, from Arctic tundra to tropical grassland. There are too few comprehensive studies even to contemplate a regional comparison. Third, sampling has often been inadequate, with only a few dates either pre- or post-fire and few seasons represented. Finally, a wide variety of sampling protocols has been used to estimate population size, including pitfall traps, sweep net samples, litter samples, soil cores, unit searching and others. These techniques are not equally good estimates of absolute population density (see Southwood 1966, Edwards and Fletcher 1971). For example, Fig. 6.1 shows a comparison of pitfall trap samples and litter volume samples extracted from data presented by Hindmarsh and Majer (1977). This comparison reveals a marked difference in the trend in total invertebrate numbers across the four sites sampled: the immediate post-fire sample yielded the greatest estimate of numbers in pitfall traps but the lowest estimate in litter samples. This result might seems to confirm the expectation that pitfall traps estimate invertebrate *activity* rather than density and may capture large numbers even if the absolute population has decreased (see below). There was also a poor relationship between sampling techniques with respect to relative abundances of the different taxa. In contrast, two other studies detected little difference between fixed-volume sampling and pitfall traps (Abbott 1984a, Abbott *et al.* 1985).

Bearing in mind the discussion in Chapter 4 about the statistical problems of assessing the effects of a fire, surprisingly few studies of invertebrate responses have sampled both control *and* impact sites before *and* after fire. Greenslade and Rosser (1984) used pitfall trapping to study litter arthropods before and after a wildfire in South Australia. These

Table 6.1. *Survey of studies of responses of soil and litter invertebrates to fires*

Reference	Methodology				Population change due to fire		
	Pre- & post-fire sampling	Control & treatment sites	Pitfall traps	Litter & soil cores	Inconclusive or none	Increase	Decline
Campbell & Tanton (1981)	x	x	x		x		
	x	x			x		
Metz & Farrier (1971)	x[1]	x		x			x
				x			x
Springett (1976)	x		x			x	
Springett (1979)	x		x			x	
Heyward & Tissot (1936)		x		x			x
Koch & Majer (1980)		x (poor)	x		x?		x?
Abbott (1984a)	x[1]	x	x		x[ab]		
	x[1]	x		x	x[ab]		
Abbott et al. (1985)		x	x		x[ab]		
		x		x			x[a]

Study					
Greenslade & Rosser (1984)		x			
Hindmarsh & Majer (1977)	x	x		x	x
Viereck & Dyrness (1979)	x	x		x	x[b]
Whelan et al. (1980)	x[b]	x		x	x[b]
Anecdotal Reports:					
Ahlgren (1974)		x			
Leonard (1974)		x			
Recher & Christensen (1981)			x		

Notes:
[1] Only a single pre-fire sample.
[a] Some taxa increased.
[b] Some taxa declined.
Source: From Tap (unpublished).

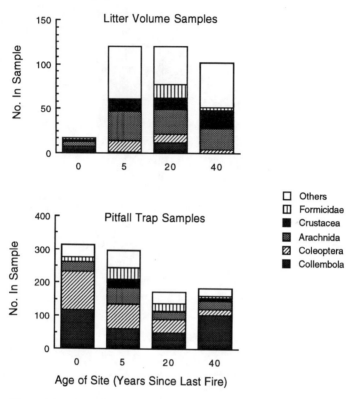

Figure 6.1 Comparison of the numbers of invertebrates in different taxa obtained in two sampling techniques: complete collection from 0.1 m² samples of litter (top) and samples from pitfall traps (bottom). Four sites in Western Australian eucalypt forest were sampled by Hindmarsh and Majer (1977), representing four ages since the last fire. The two sampling techniques produce different values for total invertebrate abundance at the four sites, and also different patterns of relative abundances of taxa across sites.

results (Fig. 6.2) illustrate the importance of establishing the pre-fire, seasonal cycle of numbers. Inspection of these graphs also indicates an increase in capture rates after wildfire, mostly for collembola and ants. Other studies have also found increased capture rates immediately after a fire (Whelan *et al.* 1980, Tap unpublished) persisting for at least a year (Fig. 6.3).

The technique of pitfall trapping does not give as good an indication of population size as sampling a known volume of litter (Campbell and Tanton 1981, Abbott 1984a). Capture rates in pitfall traps are probably more an indication of *activity* at the soil or litter surface. Thus the finding

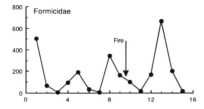
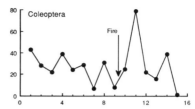

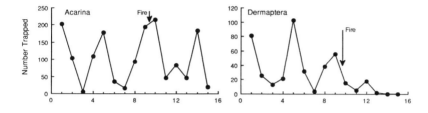

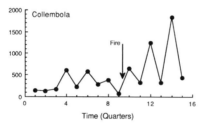

Figure 6.2 Seasonal variation in numbers of invertebrates of five taxa, sampled in pitfall traps for 24 months prior to and 18 months after a wildfire in South Australian eucalypt forest (data from Greenslade and Rosser 1984).

of increased trap rates soon after fire, using this technique, may simply reflect increased surface activity of a reduced population of animals. However, several other factors may contribute to increased trap rates and these still remain to be investigated and resolved (see Whelan *et al.* 1980). They include the following:

(i) Alteration of the complexity of the available habitat from three to two dimensions. Whelan *et al.* (1980) reported that many arthropods caught in pitfall traps after fire were taxa typical of shrub and tree foliage, not normally sampled using this technique.

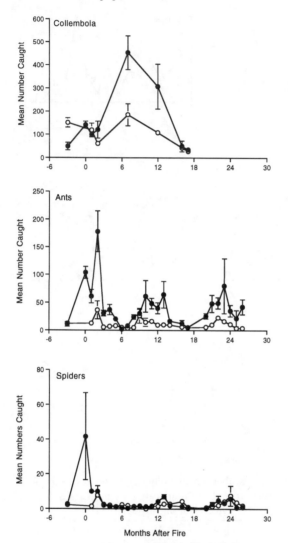

Figure 6.3 Responses of three invertebrate taxa to fire in sedgeland in eastern Australia (Tap unpublished). One of the two study plots was burned in a wildfire in January 1983 (solid symbols), after a single set of pitfall trap samples had been taken in October 1982 (− 3 months). All three taxa show an increase in capture rates following fire – ants immediately, and during each of the following two summers; spiders immediately but not in subsequent summers; and collembola delayed.

(ii) Active searching by vulnerable invertebrates for holes and tunnels to avoid predation and/or extreme physical conditions on the surface. It is not unusual to find large spiders and beetles wedged tightly, upside-down, in a test-tube pitfall trap. P. Tap (pers. commun. 1984) has even trapped frogs and lizards in this way.

(iii) Increased foraging distances or times required to maintain food input.

(iv) Lengthened foraging times permitted by changed physical conditions at the soil surface, i.e. warmer temperatures for longer time spans (see Chapter 2).

The question of whether the absolute abundance of invertebrate populations changes after fire still remains. Because there are so few adequate studies, no firm conclusions can be drawn, but the weight of a relatively large number of less-than-ideal studies (see Table 6.1) does indicate that abundances of many taxa decline after fire and may recover quite rapidly. The life-history of the organism, its habitat (i.e. is it soil or litter dwelling?) and the characteristics of the fire are likely to determine the precise nature of the decline and subsequent recovery of the population. For example, large, surface-dwelling arthropods are likely to suffer substantial mortality and reinvade from unburned vegetation, and intense fires are likely to have a stronger impact than mild ones.

One feature of some studies of soil and litter arthropod populations after fire is an apparent sustained increase in population sizes over pre-fire levels. This is illustrated clearly by the collembola and ant populations shown in Figs. 6.2 and 6.3. No explanations have been put forward for these observations, but they must reflect an increasing availability of resources caused in some way by the fire.

Given the uncertainty of the patterns outlined above, it seems unlikely that much can be contributed to the question of the impact of a fire *regime* on populations of litter and soil invertebrates. There is, however, active research on a few aspects of fire regime, principally frequency and season, focussing on the likely long-term effects of frequent, cool-season fires on the litter decomposers (McNamara 1955, Springett 1976, Hindmarsh and Majer 1977, Springett 1979, Majer 1980, Abbott 1984a, Abbott *et al.* 1985, Majer 1984). Some researchers have argued that populations of soil and litter arthropods fail to reach 'normal' (= pre-fire) levels during a 5-year inter-fire period while others contend that population levels return to normal very rapidly. Dolva (1993) found that wood crickets (Gryllidae) were more abundant in a site in south-western

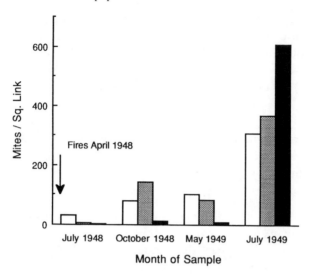

Figure 6.4 Effects of moderate-intensity (stippled bars) and high-intensity (dark bars) fires compared with no burning (open bars) on the abundance of red-legged earth mites (*Halotydeus destructor*) in pasture (data from Wallace 1961). Data are means of collections (five, 4 inch diameter cores = 62.8 in.²) from five replicate plots. High-intensity fire was achieved by the addition of dry cut grass.

Australia that had not been burned for four decades than in sites that had been burned frequently, and this appeared to be tied to the state of the litter layer.

Season of burning is likely to influence population dynamics, not only because fire intensity varies with season but also because some parts of a life cycle are certain to be more susceptible to fire than others. For example, eggs laid deep in the soil profile may be expected to survive a surface fire that was sufficiently intense to kill pre-reproductive adults. A fire of very high intensity may even kill eggs in the soil. Wallace (1961) found that an intense, late-summer fire in pasture killed mites (*Halotydeus destructor*) and their aestivating eggs. For over a year population sizes of mites remained well below the levels recorded for unburned control plots and plots burned with a moderate-intensity fire (Fig. 6.4). However, by 18 months following the fires, the population size of mites in the intense-burn plots far exceeded that in either of the other treatments.

It is strange that well-replicated field experiments have not yet been widely used to address these questions, given the suitability of invertebrates for population studies (i.e. small size, high density) and especially given their potential contribution to nutrient cycling within ecosystems, particularly in fire-prone, Mediterranean-climate regions where decomposition rates are retarded by the climate, and where fire is a common occurrence.

Macro-arthropods

Another group of organisms that should be suitable for replicated fire studies is the macro-arthropods. These animals are more mobile and are also confined to the vegetation layer in an ecosystem, so they may be expected to show different responses to fire than the soil and litter organisms. Fire temperatures are highest in this zone in most fires (see Chapter 2). Arboreal insects do not appear to survive fires well, and one study (Gandar 1982) reported almost complete destruction of the arboreal insect fauna of trees in an African savanna.

Acridid grasshoppers in south-western Australian eucalypt woodland are also removed by fire, taking some time to re-establish populations (Whelan and Main 1979). This study also illustrated the effect of one component of a fire regime, namely 'extent', on grasshopper populations (Fig. 6.5). Grasshopper populations recovered more rapidly after burns of small area than after a more extensive fire. This principle is of particular significance to understanding vegetation changes after fire (see Chapter 7).

The season of fire in relation to the organism's life cycle is also of potentially great significance. For example, Y. Gillon (1972) argued that acridid grasshopper species that lay eggs at the beginning of the dry season are most likely to experience fires early in the adult stage and are thus able to escape fire by flying and recolonize soon afterwards. Gillon listed ten grasshopper species in east African savannas that appear to have a reproductive cycle that synchronizes the early adult stage with the time of highest probability of fire. Earlier fires would kill the less-mobile nymphs and later fires would kill the less-mobile, gravid females.

Populations of vertebrates

In contrast to the common conclusion of surprisingly little animal mortality during the passage of a fire, especially for vertebrates, large post-fire changes in populations are frequently reported (e.g. Cook 1959, Arata 1959, Lawrence 1966, Bock and Bock 1978, Christensen and Kimber 1975, Recher and Christensen 1981, van Hensbergen et al. 1992, Friend 1993). Remarkable variation in responses to fire is well illustrated

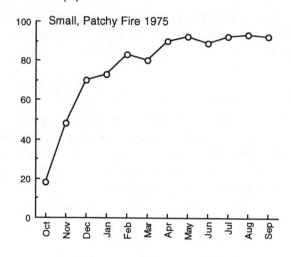

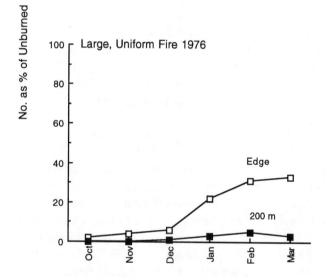

Figure 6.5 Rates of return of acridid grasshoppers to small, patchy burned areas (top) and the edge (open squares) and interior (solid squares) of a large, uniform burned area (data from Whelan and Main 1979). Fires occurred in March of each year. Data are numbers of grasshoppers captured in five, 4 m² collections (vacuumed from within a 4 m² netting cage) at each location expressed as a percentage of the numbers captured in an equivalent sample from outside the burn area, taken at about the same time.

in these mammal studies, some of which are summarized in Fig. 6.6. In north-central Florida (Arata 1959), two species of rodents showed increased abundance following fire while one (*Sigmodon*) showed a marked decline (Fig. 6.6*A*). Both males and female *Rattus fuscipes* in eastern Australia (Christensen and Kimber 1975) showed a marked decrease in population following fire, but populations in unburned sites oscillated widely (Fig. 6.6*B*). Small mammals trapped in Californian chaparral were affected by fire more severely than those in grassland (Lawrence 1966; Fig. 6.6*C*).

Studies on many species reveal an apparent decline in population sizes after fire. This is far from a generalization, however, because even within a given species there can be enormous variability both between different studies (Kikkawa *et al.* 1979) and between different areas within a single study (Newsome *et al.* 1975).

As suggested in the examples in Fig. 6.6, some studies indicate an increase in population sizes of some species. For example, populations of the introduced *Mus musculus* (house mouse) are often seen to increase after fire, from such low pre-fire numbers that they may not have been detected (Fig. 6.6*D*). The irruption of *Mus* populations appears to be a widespread phenomenon, having been reported not only in Australian heaths and forests (Recher and Christensen 1981) but also in Californian chaparral (Cook 1959), but it is not an inevitable consequence of fire (see Crowner and Barrett 1979). Furthermore, it is usually a short-term population explosion, and factors explaining the irruption and subsequent decline will be explored below. Native rodent species in both Australia (e.g. *Pseudomys novaehollandiaea*, Fox 1981) and North America, at least, also display the population growth pattern described here for the house mouse.

The reasons for the extreme variability among studies and among species in the responses of vertebrate populations to fire need to be explored further. At present it is difficult to distinguish among a large number of alternatives because most mammal studies are inadequate in some way. As with the survey of studies on populations of litter invertebrates, few studies have been able to establish the normal seasonal variation in numbers *before* applying the treatment (fire) to the site. In some cases, populations in sites that were burned appeared to be sufficiently different – before burning – from so-called control sites to be able to explain the subsequent differences between sites without invoking fire as a causal agent. In other studies, the oscillation of the population

A

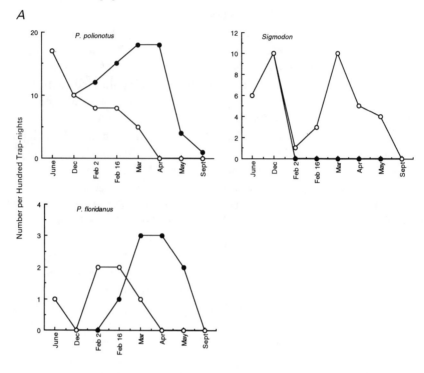

Figure 6.6 Summary of the results of several studies of the effects of fire on population sizes of small mammals. *A* Estimates of abundance of three species of small mammals (*Peromyscus polionotus* – top, *P. floridanus* – middle, *Sigmodon hispidus* – bottom) in a burned site (solid symbols) and an unburned site (open symbols) in a longleaf pine/turkey oak community in north-central Florida (data from Arata 1959). The first two data points (June and December) represent pre-burn samples. *B* Estimates of abundance of female (left) and male (right) *Rattus fuscipes*, an Australian native rodent, in a burned plot (solid symbols) and an unburned plot (open symbols) of tall open eucalypt forest in south-west Australia (data from Christensen and Kimber 1975). *C* Numbers of small mammals captured in 720 trap-nights in quarterly censuses in California grassland (left) and chaparral (right) for a burned site (solid symbols) and an unburned site (open symbols) before and after a fire (data from Lawrence 1966). *D* Numbers of *Rattus fuscipes* (a native rodent) (open bars) and *Mus musculus* (an introduced rodent) (solid bars) captured in winter samples over four years prior to and four years following an intense wildfire in heathland in south-eastern New South Wales, Australia (data from Recher and Christensen 1981).

B

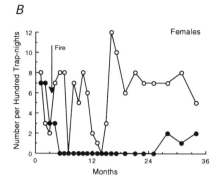

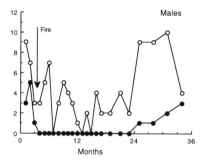

C

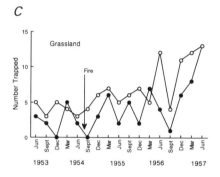

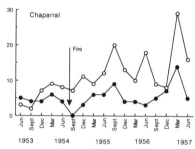

D

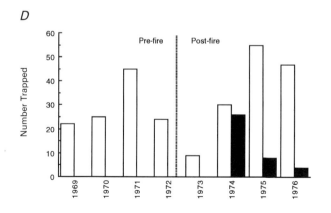

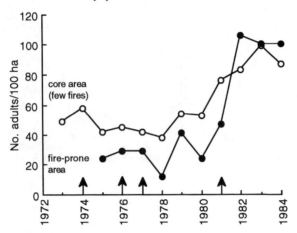

Figure 6.7 Annual censuses between 1973 and 1984 of splendid fairy wrens (*Malurus splendens*) in two parts of a study site near Perth, Western Australia (data from Rowley and Brooker 1987): the 'core area' that was not affected much by fire and a 'fire-prone' area that had suffered several fires (indicated by arrows).

curve prior to fire appears simply to be continued, again not necessarily requiring the fire in order to explain the pattern.

Studies of populations of mammals are indeed time-consuming and difficult. Nevertheless, properly designed and adequately replicated surveys are badly needed if the true effects of fire on population changes are to be revealed.

Studies of the effects of fire on the avifauna are few as well, and are mostly anecdotal (Smith 1985). However, long-term studies of bird populations by enthusiasts provide an opportunity to examine the impact of fires, and a study by Rowley and Brooker (1987) is a good example. Individuals of the splendid fairy wren (*Malurus splendens*) were colour-banded, starting in 1973, in a 120 ha study area near Perth, Western Australia. Between 1973 and 1985, six fires impinged on the study area, producing a mosaic of patches with different fire histories: some patches burned in several consecutive fires (the 'fire-prone' area) and some patches were hardly burned at all (the 'core' area). Rowley and Brooker found that occupants disappeared from territories that were completely burned in a fire and were not found in adjacent areas. However, populations built up during fire-free periods thereafter (Fig. 6.7).

In contrast to these findings after the patchy fires, a single wildfire that swept through the whole study area in January 1985 caused fewer losses of adult and juvenile birds. Rowley and Brooker proposed that one explanation of this is as follows. Most birds might survive the passage of fires. After a large fire all birds are 'in the same boat' suffering equally from the effects of the fire. After small, patch fires, however, birds in those territories that have been burned suffer badly from aggressive encouters with birds in adjacent, unburned territories.

Birds are more readily observed than many other vertebrates, and territorial species that can be colour-banded and identified frequently are good candidates for detailed studies of the direct effects of fires on vertebrates. It is surprising that these benefits have not been exploited more frequently – perhaps they should be in future.

Summary

Mortality during the passage of a fire must occur in most animal populations. However, the severity of fire-caused mortality is expected to differ with the intensity of the fire, the availability of refuges and the characteristics of the organisms. Most studies of population changes after fire are inadequate to detect more than general patterns. Populations of some animals appear to decline after fire, others appear to increase above pre-fire (or unburned control) levels. Reasons for these changes have not been clarified and are explored below.

Mechanisms of post-fire population change

Factors that might cause and maintain a post-fire decline in population (see Chapter 4) include: (i) reduced birth rates due to altered population structure (i.e. more females than males die in fire; more young animals die), reduction in food or cover, or altered physical conditions (i.e. surface maximum and minimum temperatures), (ii) increased mortality and emigration rates for similar reasons, and (iii) high predation rates.

'The greatest impediment to understanding the biotic effects of fire is the lack of data on supply of resources and levels of predation, and the paucity of field experiments' (Newsome and Catling 1983). This statement was made in relation to changes in small-mammal populations caused by fire and it remains accurate today. Few studies indeed have attempted to uncover which population parameters explain observed changes in population size or how factors such as predation, availability of resources or physical conditions affect a particular parameter. Experimental studies are possible and could be employed more widely. Crowner and Barrett (1979) conducted a field experiment enclosing populations of three species of small mammal in each of two plots, burning one plot and measuring various population parameters (sex ratio, survivorship) over time. Though this experiment suffers from lack of replication of fire and control plots, and was conducted in a grassland sown with oats, it does illustrate the sort of study that could be conducted in natural vegetation.

Newsome and Catling (1983) used a series of experimental studies investigating plagues of *Mus musculus* (e.g. Newsome 1970) to examine how changes in reproduction, mortality, age structure and total population size were correlated with five combinations of availability of food and availability of nest sites. They then compared patterns of post-fire population change in three groups of small mammals in Australian heathlands with these five models. Although the results were not clear cut, they indicated that acute food shortage may explain the patterns observed in two rodent species while predation may be implicated in the more minor changes seen in the marsupials. Shortage of shelter for nest sites did not appear to be an important factor.

Few studies have examined the effects of fire on population parameters such as age structure, size structure and reproduction. About a year after fire in south-western Australia, size distributions of two millipede species were skewed compared with unburnt, control sites, with more small and fewer large individuals present (Springett 1976). Because absolute abundances appeared to decline in response to fire, Springett interpreted the cause of this altered size structure as either increased birth rates or increased survival rates of young individuals in burned vegetation.

After studying small mammals in Californian chaparral, Lawrence (1966) commented that mortality due to fire brought about a shift in the age structure of *Peromyscus truei* populations, resulting in a greater proportion of young animals, and a consequent stimulation of the reproductive rate. In contrast, Wirtz (1977) reported reduced reproductive success in raptors after chaparral fire. Christian (1977) studied two nocturnal rodent species in south-west Africa (*Desmodillus auricallaris* and *Gerbillus paeba*). For animals in a burned and an unburned site, there were no significant differences in measures such as mean survival rate, reproductive rate or timing of breeding.

Table 6.2. *Causes of mortality of woylies (*Bettongia penicillata*) and tammars (*Macropus eugenii*) in a study of the effects of fire on these species in south-west Australian eucalypt forests*

| | Deaths recorded | | | |
| | Betongia | | Macropus | |
Probable cause of death	No.	%	No.	%
European fox predation	15	45.4	7	17.1
Native cat predation	7	21.2	0	0
Eagle predation	1	3.0	5	12.2
'Fire shock'[a]	2	6.1	0	0
Fire	1	3.0	0	0
Paralysis[b]	2	6.1	24	58.5
Other	5	15.2	4	9.8
Starvation	0	0	1	2.4
Totals	33	100	41	100

Notes:
[a] appeared to be some sort of psychological shock brought about by the after-effects of the fire, rather than the fire itself.
[b] associated with trapping.
Source: From Christensen (1980).

Post-fire food availability

The study by Christensen (1980), outlined above, recorded 74 woylie (*Bettongia penicillata*) and tammar (*Macropus eugenii*) deaths during a study in south-west Australian eucalypt forest. Only one of these deaths was assessed as having been caused by starvation in a recent burn area (Table 6.2). This was perhaps a surprising finding, because woylies, in particular, adhered strongly to their home ranges after the intense, experimental fire that burned their habitat. An explanation for the lack of starvation might be the fact that the woylies exploited a hypogean fungus (*Mesophellia*) as a food source. Fire apparently makes this resource more available, and woylie scratchings with *Mesophellia* remains were observed 30 to 70 times more frequently in burnt than in unburnt areas. Berenstain (1986) reported a shift in food resources used by several primate species in eastern Borneo during the drought in 1982 and 1983, and the fires that followed. The animals under study (long-tailed macaques, gibbons, orangutans and gibbons) tolerated the fire and no changes in group composition were recorded. Reproduction also occurred in the post-fire period. During this time, however, the macaques showed a marked shift in diet, feeding mostly on foliar and herbaceous material and insects such as caterpillars and larvae of wood-boring beetles. The preferred items, flowers and fruits, were unavaliable following the fire.

The conclusion that fires have long been an integral part of the environment for many animals would lead us to expect mechanisms, such as the feeding specialization described for the woylie, that permit survival in post-fire conditions. However, few have been clearly revealed. In fact, avoidance of post-fire starvation is by no means a general observation. Bigalke and Willan (1984), for example, suggested that the disappearance of herbivorous rodents (*Otomys* spp.) from burned areas in South Africa is caused directly by food shortage. Bock and Bock (1978) found that populations of the cotton rat (*Sigmodon hispidus*) in Arizona grassland were greatly reduced by a summer fire while the heteromyid rodents (*Perognathus hispidus* and *Dipodomys merriami*) increased. These differences were attributed to the food habits of the animals, with cotton rats feeding on green vegetation, which declined after fire, and the other species feeding on seeds, which increased in abundance due to post-fire invasion of weedy forbs.

In a more direct indication of the effects of food availability, Newsome et al. (1975) showed that the mean weights of small mammals surviving an intense fire in heathland in south-eastern Australia were lower than those for the populations prior to the fire (Fig. 6.8). Although this difference in population means could be attributed to an altered age-structure (i.e. high proportion of young, smaller animals), 35.2% of 91 *Rattus fuscipes* individuals lost weight over winter in the year after fire compared to only 16.9% ($n = 219$) losing weight over the winter before the fire. Lawrence (1966) also reported a reduction in mean weight of a rodent population (*Peromyscus truei*) after fire in Californian chaparral and attributed this partially to a shift in the age structure of the population because resident adults were killed and their places taken by smaller juveniles. Notwithstanding this explanation, it was apparent from Lawrence's study that individual animals suffered from under-nourishment in the first years after fire. This was assessed by calculating the ratio of body weight to body length as a condition index. Such indices have been used in many animals to relate nutritional status of the animal to environmental variables such as drought and seasonal changes (Murie 1963). The condition index for *P. truei* exhibits normal seasonal

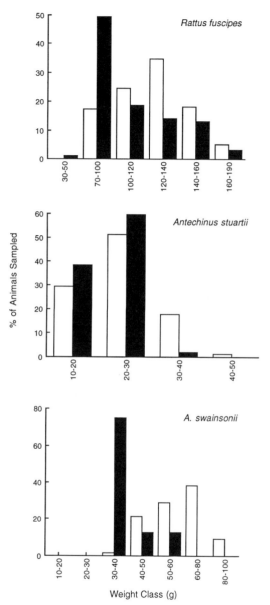

Figure 6.8 Different weight–frequency distributions for three species of small mammals (a rodent, *Rattus fuscipes* (top), and two dasyurid marsupials, *Antechinus stuartii* (middle) and *A. swainsonii* (bottom)) recorded during April–November in the year prior to (open bars) and after (solid bars) a severe wildfire in heathland in south-east New South Wales, Australia (data from Newsome *et al.* 1975).

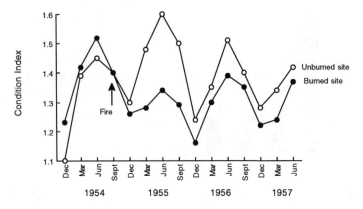

Figure 6.9 Seasonal changes in condition (body length (cm)/body weight (g)) for the rodent, *Peromyscus truei*, captured in a burned site (solid symbols) and an unburned site (open symbols) both before and after an experimental fire in Californian chaparral (data from Lawrence 1966).

variation with a peak in condition occurring in summer (June), but this peak is not reached for animals in burned chaparral for about two years after the fire (Fig. 6.9). In a study of the effects of fire on a population of snowshoe hares (*Lepus americanus*) in Alberta, Canada, Keith and Surrendi (1971) found that the pregnancy rate for females living in a burned area (42%) was significantly lower than for animals in unburned areas (76%).

The examples of declining abundances of invertebrate and small mammal populations after fire, mentioned above, would indicate an absolute shortage of food for predator populations even if there is a *relative* increase in prey due to increased availability. There are few data describing the longer-term effects of fire on predators. Newsome *et al.* (1975) reported a change in the diet of dingoes from small mammals before fire to wallabies and kangaroos after fire in Australian heathland. Wirtz (1977) reported a decline in the reproductive success of several raptor species after fire in Californian chaparral. In contrast to this finding, Lawrence (1966) found a small increase in abundance of predatory mammals in the year after fire in Californian chaparral, and also a sharp increase in the abundance of predatory birds that lasted for several years after a fire.

An *increase* in food availability after fire is suggested by the examples of population increases mentioned above. One possibility here is that a fire favours food–plant species that were previously rare and therefore permits access to those animal populations able to use the increased resource. Thus, the post-fire increase in population sizes of rodent species such as *Reithrodontomys* in Californian chaparral is probably related to the increased availability of seeds of annual grasses stimulated by fire (Cook 1959, Ahlgren 1966). An alternative explanation for increases in populations is a sustained increase in food availability without a shift in species of prey. This suggestion is supported by many reports of increased forage *quality* for herbivores after fire. Prescribed burning is associated with increases in the protein and phosphorus content of new growth (e.g. Dewitt and Derby 1955), and foliage texture also changes over time after fire. New shoots soon lose their softness and succulence and become tougher and therefore less nutritious (see Hilmon and Lewis 1962, Hobbs and Spowart 1984). Using tame animals, Hobbs and Spowart (1984) conducted some elegant experiments to examine the effects of fire on the quality of diets selected by ungulates in Colorado rangelands. Fire produced significant increases in forage quality for both mountain sheep and mule deer (Fig. 6.10). Do these changes in food quality result in quantifiable increases in population growth rates? This seems likely, and studies of sheep grazing on burned heather in Ireland have shown an increase of 15% in lambing, an increase of 32% in live birth weight per lamb and an increase of 30% in lamb growth rates (Lance 1983). Studies such as these are badly needed for native herbivores in ecosystems subject to natural fires.

Circumstantial evidence for the importance of improved forage quality is provided by many observations of herbivore populations concentrating in recently burned areas (Phillips 1965, Frost 1984, Noble *et al.* 1985). Movement into recently burned areas is not a universal finding (Caughley *et al.* 1985) but it appears to be most commonly reported in highly mobile mammals such as macropod marsupials in Australia or ungulates in Africa (Phillips 1965, Oliver *et al.* 1978) and North America (Daubenmire 1968). These animals may simply move into favourable areas as they become available and it may be difficult to demonstrate any difference in population growth parameters between animal populations in a broad landscape containing burned areas and populations in areas unaffected by recent fire. However, for nearly a century, management of heather (*Calluna vulgaris*) moors in Scotland to maximize population sizes of grouse has included burning in small patches so that individual animals are able to use both mature heather, for cover, and recently burned heather, as a high-quality food (Lovat 1911). Picozzi (1968) found a strong correlation between the number of grouse

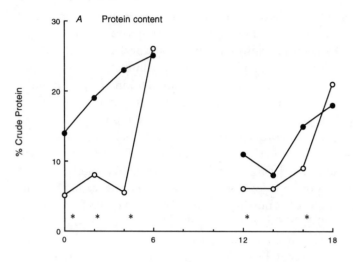

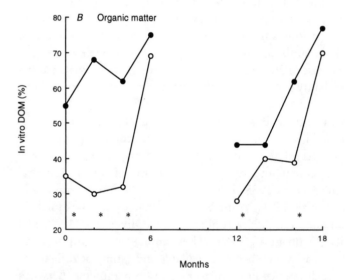

Figure 6.10 Crude protein content (*A*), and *in vitro* digestible organic matter (*B*) in diets selected by mule deer over two winters in replicated burned (solid symbols) and unburned plots (open symbols). Asterisks represent statistically significant differences (modified from Hobbs and Spowart 1984).

shot per km² per yr and the number of recently burned patches of heather traversed in transect surveys of 26 estates.

Increases in food availability have been suggested as reasons for increases in bird abundance in recently burned sites, particularly of seed-eating birds that forage on or near the ground (Recher *et al.* 1985, Christensen and Kimber 1975). However, in each of these studies, it is likely that the birds moved into the burned area from outside, and increased growth rates, breeding success and recruitment of resident birds has yet to be demonstrated. Nevertheless, vegetation recovery is clearly important for the recovery of bird populations, especially of species that inhabit the ground layer and shrub vegetation (Smith 1985, Recher *et al.* 1987), and this regenerating vegetation must be providing food, nest sites and protection from predation.

Post-fire mortality by predation

The potential severity of post-fire predation is illustrated by frequent observations of insectivorous birds gathering at fires (Komarek 1969, Y. Gillon 1972, Frost 1984, Kikkawa *et al.* 1981). Gillon recorded large numbers of grasshoppers, crickets and mantids in the stomachs of 14 raptorial birds sampled while feeding in smoke plumes (Fig. 6.11). Although the arthropod fauna in the savanna included nymphs and flightless species, all insects in the stomach contents were strong fliers, implying that they had been captured flying in front of the flames.

Increases in predator populations may last for some time after a fire, as indicated by the findings of Lawrence (1966). Numbers of breeding predatory birds in his 20 acre experimental plot increased from 1.3 prior to the fire to 9, 5 and 4 during the following three years. Similarly, Bock and Bock (1978) reported an increase in raptor populations that lasted for a year after fire in grassland.

Numerous studies mention predation as a major source of post-fire mortality of mammals (e.g. Christensen and Kimber 1975, Newsome *et al.* 1975, Bigalke and Willan 1984, Christensen 1978; Table 6.2), but few have attempted to quantify its effects on the population structure and dynamics of the prey species. Only inferences are possible. For example, Christian (1977), in a study of the responses to fire of south-west African desert rodents, noted that the one rodent species that was adversely affected by fire (*Rhabdomys pumilio*) is a diurnal species and appears to require a dense bush and grass canopy for cover.

Intense predation by insectivores on prey made more available by fire

Figure 6.11 Concentration of kites feeding on insects fleeing the fire front in a January fire in west African savanna. (Photo by Y. Gillon 1972, reproduced with permission of the author and the publishers, Tall Timbers Research Station.)

may explain some of the variability among studies of leaf litter invertebrates described above. Pitfall trap samples taken immediately after fire have recorded great surface activity (Whelan *et al.* 1980, Tap and Whelan 1984), whereas other studies using pitfall traps some time after fires, perhaps when predation has reduced population sizes, have recorded reduced capture rates for various taxa (e.g. Abbott 1984a, Abbott *et al.* 1985).

The potential importance of predation as a source of post-fire mortality is also suggested by observations of colour change in arthropod populations in burned areas (Burtt 1951, Hocking 1964). Hocking's study, in the Sudan, showed that geophilous (ground-dwelling) grass-hopper species tended to be dark in colour while phytophilous species were mostly pale. Moreover, several species display intra-specific colour variations. Further work should be conducted on this interesting subject. It is not clear, for example, whether paler individuals select paler backgrounds, or what cues are used to detect a suitable background. Also uncertain is the mechanism for a colour change within a population after fire. Simple selection by predators could account for it, as in the case of industrial melanisms. However, at a moult, formerly light-coloured nymphs could take on the darker shade of the burned background, and Burtt (1951) demonstrated experimentally that adult grasshoppers (*Phorenula werneriana*) were able to take on the colour of various backgrounds, including burnt ground. Finally, the actual advantage of being dark on a burnt background need not solely be escape from the attention of predators. Increased activity due to enchanced radiant heat absorption is another possibility (Frost 1984).

From the point of view of the predators, food may be made more available by fire, because prey are flushed from habitats within which they are usually cryptic. For example, many of the invertebrates trapped after fire in pitfall traps at ground level are representatives of groups normally confined to shrub and tree canopies. Although these indivi-duals survive the passage of the fire front, the consumption of above-ground vegetation by fire confines them to moving on the ground. Many groups that are cryptic against the green background of shrub foliage become highly conspicuous against the blackened soil surface (Whelan *et al.* 1980). If prey are forced to forage more widely, for longer periods and without cover after fire, they are likely to be an *available* resource for some time. The altered vegetation structure may also make a higher proportion of a habitat accessible to predators. Foxes are now common predators in Australian heaths and forests, but they forage preferentially along existing tracks and trails and not in dense under-growth (N. H. Robinson, pers. commun. 1986). Fire would therefore permit more widespread hunting until a dense vegetation cover had regenerated.

Explaining different patterns of population response

The above discussions illustrate enormous variation among animal species in their population responses to fires. Some invertebrate groups decline in abundance after a fire while others display a fire-related increase. Similarly, some bird populations increase and others decrease as a result of fires. Populations of native rodents disappeared from some areas of burned heath in south-eastern Australia, while marsupial populations persisted in the same area. Explanations of population responses to fire must take variability like this into account.

Variable responses among species immediately after fire may be explained by the life-history characteristics of the animals and their habitats. Animals living in protected habitats are more likely to survive than those in exposed areas. For example, sedentary, foliage-dwelling arthropods are less likely to tolerate the passage of fire than burrowing, soil invertebrates. Similarly, small mammals living in moister habitats such as creek margins may be more likely to survive than individuals on dry ridges. In the longer term, characteristics of an animal species such as mobility and particular food and cover requirements will determine its ability to re-invade a burnt site.

The characteristics of a fire interact with these features of animals to determine survival and re-establishment of a population. For example, intense, fast-moving fires may cause greater mortality of surface- and foliage-dwelling animals than cool, slow-moving fires. A fire in one season will have a different effect on a population than a fire at some other time of year if different life-history stages are affected. Good examples of this might be bird populations that are more susceptible to fire at nesting time and insect populations that are more resilient if fire occurs after eggs have been deposited in the soil.

Warren et al. (1987) reviewed the complex web of interacting effects of fire on arthropods in grasslands, and highlighted the various factors contributing to population changes in response to fire (Fig. 6.12). They separated the effects of fire into four 'phases' (shown along the bottom of Fig. 6.12): (i) pre-burn conditions, (ii) combustion phase, (iii) immediate post-burn 'shock' phase and (iv) longer-term recovery phase. During and after a fire, different components of the invertebrate community are likely to be affected differently, and three main classes of 'response groups' are recognised as separate boxes in each phase: (i) surface-dwellers/flying, (ii) surface-dwellers/non-flying, (iii) soil-dwellers. Different life-stages of a particular organism may be also be affected differently. Factors reducing populations at each phase are indicated by downward-pointing arrows: various causes of mortality and emigration. The principal environmental factors determining the rate of transition from one phase to the next are indicated between the boxes representing the invertebrate community. This pictorial model serves as

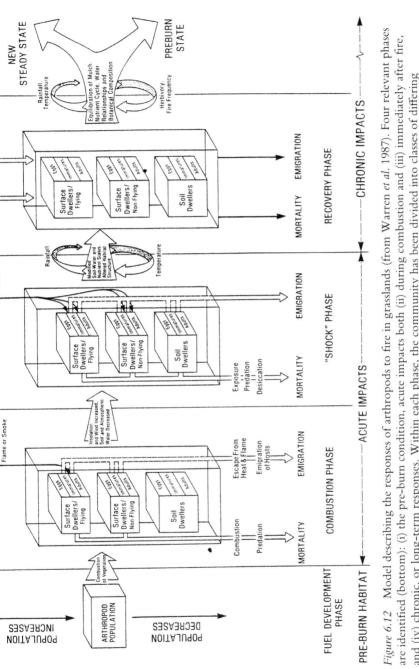

Figure 6.12 Model describing the responses of arthropods to fire in grasslands (from Warren *et al.* 1987). Four relevant phases are identified (bottom): (i) the pre-burn condition, acute impacts both (ii) during combustion and (iii) immediately after fire, and (iv) chronic, or long-term responses. Within each phase, the community has been divided into classes of differing microhabitat and habit (surface-dwellers/flying, surface-dwellers/non-flying, soil dwellers), and within each of these, different life-history stages are recognized. Arrows coming into the 'community boxes' from above indicate sources and magnitudes of inputs to populations/life-history stages and arrows pointing down from the boxes indicate sources and magnitudes of losses to populations.

a good 'checklist' of processes, affected by fire, which determine population dynamics.

One feature of fire regime that is of great importance in survival and re-establishment of animal populations is patchiness. Animals have a great advantage over plants in that they are generally mobile and are therefore able to move to avoid the direct effects of radiant heat in a fire (Chapter 3). The potential importance of escape in animal populations confronted by fire was well illustrated in the studies of acridid grasshoppers in an Ivory Coast savanna (Y. Gillon 1972) described above. Such mobility allows animals to exploit refugia, that is, particular locations that escape being burned for one reason or another.

Lawrence's (1966) study of small mammal populations after fire in Californian chaparral showed that patches of outcropping rocks within the chaparral provided locations in which small mammals were little affected by fire. Various other studies have also suggested that unburned refuges provide the nucleus for recovery of animal populations after fire (e.g. Recher and Christensen 1981). However, despite these observations, and the logical predictions that follow, few direct studies have been conducted. Do fleeing animals congregate in patches of unburnt vegetation simply because they move on every time the local patch starts to burn? Do they actively seek out patches of vegetation that are unlikely to burn? Are they rejected by the normal residents of the patch? One of the few pieces of information relevant to these questions was obtained by Gandar (1982) in a study of grasshopper abundances in burned and unburned patches of savanna in Africa. Density of grasshoppers increased markedly in the unburned patches of vegetation and subsequently declined as animals reinvaded the recovering, burnt vegetation (Fig. 6.13). In another contribution to this question, Whelan et al. (1980) noted that many arthropods could be found alive in the dense crowns of rosette-like plants such as Xanthorrhoea and Macrozamia in the first few days after fire (see Table 3.5). The growth form of these plants provides a microsite that escapes the intense heat of the fire (e.g. Koch and Bell 1980). A greater number of individuals and a greater variety of arthropod species were found in these microsites in the midst of a burned area than in equivalent plants in unburned areas.

It is important that studies are initiated that will develop our understanding of the responses of herbivore populations to post-fire vegetation changes, especially in relation to fire patchiness, because these factors may have a strong influence on the responses of plant populations to fire (Friend 1993). This will be investigated further in Chapter 7.

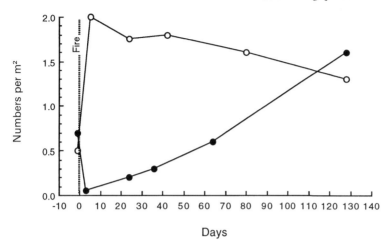

Figure 6.13 Population densities of grasshoppers in burned (solid symbols) and unburned (open symbols) patches of African savanna both before and after fire (from Gandar 1982 and Frost 1984). The large increase of animals caught in unburned patches immediately after the fire, followed by a steady decline, mirrored by opposite patterns in the burned sites, suggests an escape to unburned refugia followed by post-fire reinvasion.

Outstanding questions

What is most needed in the field of responses of animal populations to fire is a properly designed and analysed assessment of population dynamics during and following a single (well-described) fire for any animal species. Such a study would generate the relevant questions regarding the relative importance of cover, predation and food in regulating population size after fire. Furthermore, only such a study will refine the questions that need to be asked about a fire *regime*, rather than about the effects of a single fire. Of course, long-term experiments designed to assess the effects of fire regime on population dynamics of a whole range of animal groups should also be established. These will require permanent marking and frequent censusing of individuals: a major challenge for future fire ecology. Since it is difficult to predict where and when wildfires will strike, long-term studies with regular censuses of marked animals are badly needed (e.g. Rowley and Brooker 1987).

In addition to these general research needs, the following questions appear to be particularly interesting and/or important.

1. Studies on the responses of animal populations to fire could focus in more detail on the question of to what degree different sorts of fires cause mortality, emigration and survival within a site. We might expect that factors such as patchiness of the fire, continuity of the fire front, and intensity and season of the fire will all be relevant factors.

2. Related to this question is the need to investigate modes of recovery of populations on burned sites: i.e. by recolonization from boundaries or from within the burn boundary.

3. In terms of animal survival of fire in unburned refuges, it is important to know whether the organisms found in refuges after fire represent just those individuals that happened to be present prior to the fire or else an increased density due to active seeking out of areas that escape being burned.

4. The question of whether mortality is insignificant only in the more frequently studied prescribed or 'controlled' burn needs further investigation. If wildfires exact a greater toll on animal populations, especially of herbivores, there could be serious consequences for plant populations and communities of using prescribed fire to reduce wildfires over large expanses of country.

5. The relative importance of food, cover and predation as factors controlling post-fire animal populations requires further investigation. Elegant experiments are possible here, as factors such as food, cover and possible predators could be experimentally manipulated (e.g. Dickman 1989, Broughton and Dickman 1991) in replicated burned and unburned enclosures.

6. The apparent contradictions regarding the direction of population changes of soil- and litter-dwelling invertebrates after fire requires closer examination and carefully designed experiments.

7 · *Community responses to fire*

Introduction

The component of an ecosystem that is most readily perceived in relation to fire is the community. A casual observer of a recently burned area registers the apparent absence of herbs and shrubs, the blackened remains of trees, carcases of animals, without necessarily contemplating the responses of particular species or likely changes in population size.

Except in the case of endangered or unusual species that have already been identified by a relevant land manager as important, the community is also the level usually addressed by management, with the aim of maintaining species diversity, dominance, or a particular assemblage of species (e.g. van Wilgen *et al.* 1990, van Wilgen *et al.* 1992a).

The community is also the biotic unit that receives a fire. There is therefore a clear interaction among species. The nature of the fire experienced by a particular individual plant or population of plants, for example, will be determined partially by the characteristics of the surrounding plants, by factors such as live and dead biomass, relative abundances of species of differing flammabilities, and vertical and horizontal structure. Similarly, the fates of many of the animal species present in an area, both during and after a fire, will be dependent upon the plant component of the community for various reasons – food species present or absent, the cover provided, the type of fire sustained, etc. It is therefore clearly an unreal situation to view individual organisms or populations in isolation, when drawing conclusions about the effects of fire in nature. This more detailed approach, focussing on individual organisms and on populations, is certainly necessary but it must eventually be put back into the context of the community.

The term 'community' triggers a range of perceptions. It therefore requires some definition here. Further elaboration and discussion of the definitions are found in texts such as Krebs (1985), Begon *et al.* (1986), Diamond and Case (1986), Kikkawa and Anderson (1986), and Gee and Giller (1987). A community is defined here as a group of co-occurring species that are potentially interacting. This broad definition therefore includes in a community many taxa and many trophic levels (see Table 7.1). It must be recognized that, for convenience, the term community is often applied to a subset of the taxa present in an ecosystem – most often a 'plant community'. This taxonomic bias has contributed to a general lack of attention paid to the role of interactions, especially a lack of attention in the botanical literature to the role herbivores. In relation to fire, the importance of herbivory in altering the composition of the plant component of a community can be clearly demonstrated using experimental exclosures. Studies incorporating other important groups of organisms, for example insect herbivores, symbiotic bacteria and fungi, will undoubtedly be more difficult and complicated, but are increasingly being recognized as important.

In relation to plants, when defining a community as a group of co-occurring species that are potentially interacting, it is not assumed that a community is a discrete unit of organization rather than simply an assemblage of populations that happen to co-occur. This has been discussed recently by Austin (1985, 1990), Noy-Meir and van der Maarel (1987) and others. It is an important discussion, because fire-induced changes in the presence and absence of a few plant species are often described as shifts from one 'plant community' to another. Of course, in viewing a community in a broad sense, including a range of taxa and trophic levels, it is clear that some relationships will be more than just co-occurrence. Specialist herbivores will be associated with just one or a few plant species; pollinators may be associated with particular suites of plant species; some vertebrates will depend upon certain vegetation structures for cover and nest sites.

There have been two main approaches to the study of community ecology. One could be called 'reductionist' because of a focus on the detail of interactions between populations of organisms, aiming to build up an explanation of community composition and changes in it from an understanding of how the component parts (populations) influence each other. Many of these population interactions have been dealt with in previous chapters.

The other approach may be called 'holistic' because it focusses on the changes occurring in some emergent property at the community level: some indicator of composition or structure, such as species diversity, dominance or trophic structure. Further than this, the holistic approach includes measures such as living and dead biomass, percentage ground cover – factors that contribute to the behaviour of a fire even though

Table 7.1. *(a) Definition of 'community' and its various components*[a]*. (b) Summary of range of applications of the term 'community' to various groupings of species*[b]

(a) Level	Patterns	Definition
Community	Trophic structure Competitive interactions Taxonomic diversity	Group of organisms, generally of wide taxonomic affinities, occurring together. Many interact with horizontal (i.e. competition) and vertical (i.e. predation) linkages.
Sub-community	Habitat preference Resource partitioning	Taxonomically or functionally restricted group of species occurring together. Interactions mainly by horizontal linkages
Guild	Species size patterns Resource preferences	Groups of ≥ 2 co-occurring populations using similar resources in similar ways. Interactions mainly by horizontal linkages.

(b) Class of definition	Definition
Locational	All living organisms in habitat or prescribed area Any organism in stratum of habitat Any assemblage of organisms in a prescribed area Groups of species living closely enough to interact
Trophic	Species in all trophic levels All species in adjacent trophic levels All species in a single trophic level All species with similar resource requirements Species using similar resources in similar ways
Taxonomic	All species in a prescribed area Large-scale taxonomic assemblages All species in a restricted set of taxa Species in one taxon, with similar resource requirements All species with a particular life-form

Sources: [a] Modified from Table 23.4 of Gee and Giller (1987). [b] From Table 23.3 of Gee and Giller (1987).

they may not necessarily contribute information directly on biotic interactions within the community.

Addressing the question of how a community responds to fires may require a combination of the holistic and the reductionist approaches. So little is known about the effects of fire in most regions that description of changes, carefully related to characteristics of the fires, is still badly needed. It is not profitable to attempt such a description separately for every species, so measurement of changes in species richness, diversity, biomass, percentage cover, dominance or other community-level parameters may be an effective way in which to compare the effects of different fire regimes. Understanding *why* the observed changes have taken place will generally require much more detailed, experimental studies of the effects of physical factors and of specific interactions between populations, focussing on processes such as competition, herbivory, predation and mutualisms.

Direct measures of community attributes

Effects of a single fire

Various categories of response of plant species to fire were listed in Chapter 3 (see Table 3.4). This classification was based largely on life-history characteristics and on the responses of individual plants. Jarrett and Petrie (1929) and Purdie and Slatyer (1976) have used slightly different classifications, based on the population responses. Defining population response categories allows direct inferences to be made about the changes in community structure that are likely to occur as a result of fires. For example, the term 'fire-sensitive increasers' defined a set of species in which established plants may die but the population nevertheless increases after a fire, because of good germination and dense seedling establishment. In contrast, the same fire can produce a decline in population size for the 'fire-sensitive decreasers'. For a given fire, classification of species into these two groups allows some level of prediction about how a plant community might change in dominance, relative abundance and relative cover as a result of a fire.

Uncritical adoption of classification schemes such as this one was criticised in earlier chapters. Just as established plants of a species may sprout after fire in one region and die after a similar fire in another, or sprout after a low-intensity fire and not after a high-intensity fire (Trabaud 1987; see also Chapter 3), so populations are likely to change in different ways after different fires. Hence a 'fire-sensitive increaser' identified after a high-intensity fire with good follow-up rains may become a 'fire-sensitive decreaser' as a result of a subsequent fire with poor following rains, or higher levels of post-fire herbivory, or lower-intensity fire that fails to stimulate good germination. Similarly, Rowe (1983) identified the fact that sprouting of shrubs after fire will depend upon the combination, at a particular site, of depth of meristems under the soil and intensity of the fire at ground level. Thus *Vaccinium vitis-idaea* may have shallow, subterranean runners (in Sweden; Uggla 1958), it may be rooted in the moss layer (in Alaska; Chapin and van Cleve 1981) or it may be deep-rooted (in Manitoba, Canada; Ritchie 1959). Accordingly, it would be classified as intolerant of fire in the first two reports and an 'endurer' (Rowe 1983) in the last. In either sprouters or obligate seeders, more subtle variation in species composition follows fires. Moreno and Oechel (1991) showed that different fire intensities in Californian chaparral can have profound effects on species composition, through the different germination responses to different fire intensities of some common species of shrubs and herbs. The important point is that we can be certain that a community will change in different ways depending upon the characteristics of the fire itself, the past fire regime, and post-fire climatic and biotic interactions.

The following discussion examines various patterns of community responses to fire. It will be apparent that an analysis of community responses to fire must come from a combination of synchronic and diachronic studies; that is, instantaneous comparisons of many sites of different post-fire age combined with longitudinal studies in a set of sites for long time periods.

Changes in species richness – plants
Descriptions of classical secondary successions (e.g. old fields) typically feature a rapid decline in species richness as the disturbance eliminates many intolerant species, then an increase in richness as species reinvade and establish, often followed by a slow decline as the plant community matures (see Whittaker 1975, Odum 1969). The generality of this pattern of communty change, at least in north-eastern US old-field successions, led to an expectation that this will also occur after fires (e.g. Jarrett and Petrie 1929, Shafi and Yarranton 1973). Thus, early descriptions of secondary succession included fire as one of several possible agents of disturbance seen to produce the 'typical pattern'.

Dry-land successions can be illustrated for the oak–pine forests of Long Island, New York, which have been subject to extensive clearing for farms and to frequent fires. After either destruction of the forest by fire or abandonment of a farm field, a succession leads back to forest again . . . There are differences between successions following fire and those following farm abandonment; but the general pattern may be described by stages . . .   (*Whittaker, 1975; pp. 174–5*)

The 'typical' pattern of changing species diversity certainly does appear to hold after fire in some communities, whenever a fire removes all vegetative and seed material from a wide range of species, and reappearance of photosynthesizing plants depends on their seeds dispersing into the site and seedlings being able to establish once there. Detailed observations by early natural historians of the development of a flora after fire are well illustrated by an account (Colgan 1913) of the origin of plants appearing after a fire near Dublin. Sixty four species of higher plants and cryptogams were found in a burnt area within 18 months of the fire and, of these, at least 45 were considered to be immigrants to the site. This group was 'made up chiefly of species provided with special contrivances for seed dissemination, or with very light or minute seeds or spores adapted to wind carriage'. Similarly, studies on the effects of fire on forest and forest–tundra in high latitudes such as Alaska, Canada and Japan (Lutz 1956, Wein and MacLean 1983, Nakagoshi *et al.* 1987) have revealed that fires are likely to eliminate many plant species occupying a site, reducing species diversity and making the site accessible to a sequence of invading species that were not locally present prior to the fire.

A superficial examination of change in plant species richness after fire in other, more fire-prone regions of the world, such as the fynbos of South Africa, the chaparral of California and Australian heaths and forests, may reveal some similarities (Figs. 7.1 and 7.2) with the above patterns. Studies by Bell and Koch (1980), Posamentier *et al.* (1981), Hobbs and Atkins (1990), Specht *et al.* (1958), Adamson (1935), Trabaud and Lepart (1980) and Vogl and Schorr (1972), for example, can all be interpreted as showing an increase in species richness reaching a peak and then declining over relatively long time periods.

A more thorough analysis reveals, first, that the initial increase in richness observed after fire in many ecosystems is contributed to almost exclusively by species recovering from organs or seeds *on site*. A drop in number of species from pre-fire levels is because the fire removes all easily observed plant material. These species are subsequently recorded again as seed germinates or sprouting occurs. Second, although a peak in species richness may be perceived in some studies (e.g. Bell and Koch 1980, Hobbs and Atkins 1990, Trabaud and Lepart 1980), it need not reflect 'a sere with an admixture of pioneer and climax species' (Bell and Koch 1980). An increase in species richness soon after fire may simply reflect the fact that some species in the pre-fire community were represented as seeds or dormant underground organs, and did not, therefore, contribute to censuses at that time.

These possibilities reveal a difficulty in the interpretation of many studies of changes in species richness as a result of fire: just what is defined as the plant community in which these measures are being made? If above-ground, living, plant material is counted as the sole contributor to species richness, then an intense fire can reduce richness temporarily to zero. At the other extreme, if seeds and surviving below-ground plant parts are included, species richness may remain completely unaltered by fire in some locations. The distinction between these two approaches to measuring richness is important in the context of interpreting post-fire changes in a plant community as successional.

One other important question requires some careful consideration here: how is species richness or diversity to be measured? Studies of plant community change after a fire have typically been assessed using quadrats of one size or another: $1.6 \, m^2$ and $2.0 \, m^2$ (made up of 16, and 20, $0.1 \, m^2$ respectively; Posamentier *et al.* 1981), $4 \, m^2$ (Keeley and Keeley 1984), $100 \, m^2$ (Trabaud and Lepart 1980); $400 \, m^2$ (Fox and McKay 1981). Using 25, $1 \, m^2$ quadrats nested in 5 m squares, Muston (1987) showed that the pattern of changes in species richness after fire in an eastern Australian eucalypt forest depends upon quadrat size. An apparent decline in species richness after fire could therefore be interpreted as an artifact of changes in species *density* (Fig. 7.3). This is easily understood, because as seedlings become established and sprouts grow, the space available in a $1 \, m^2$ quadrat is quickly occupied. Further growth, random mortality and density-dependent thinning will then see a reduction in the numbers of plants in each $1 \, m^2$ and hence in the mean number of species per quadrat.

Changes in species richness – animals
Changes in animal species richness in relation to fire are perhaps best known for small mammals and birds. Changes in the animal component of a community certainly occur after fire, with some species disappearing, and then reappearance of species at different times during vegetation recovery. As with plants, these processes of species disappearances and reappearances produce patterns of change in species diversity (Fig. 7.4), and these patterns include, in various studies, increase in richness up to a

A

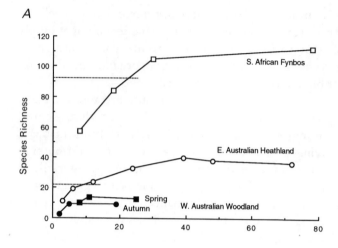

B

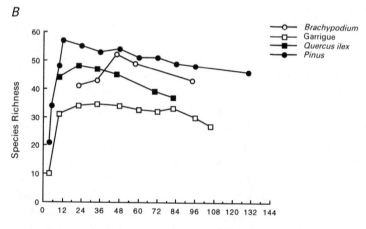

Figure 7.1 Summary of changes in plant species richness over time since fire, in diachronic studies (single sites sampled over time after fire).
A. Comparison of sites in South African fynbos (Adamson 1935, after Kruger 1983), south-eastern Australian heathland (from Posamentier *et al.* 1981) and western Australian *Banksia* woodland burned in spring and autumn. Dashed lines represent pre-fire measures of species richness in fynbos and eastern Australian heathland (from Hobbs and Atkins 1990).
B. Four plant associations in Mediterranean garrigue (from Trabaud and Lepart 1980).
C & D. Six plant associations in south-eastern Australian eucalypt forest (from Muston 1987).

C

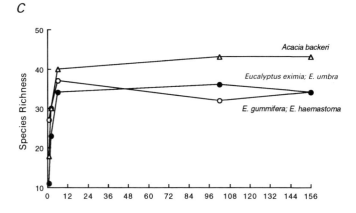

D

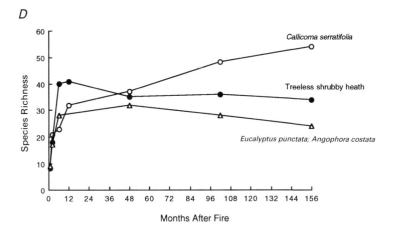

Months After Fire

constant level and also an initial increase followed by a decline. Change, or constancy, in species richness can, of course, obscure important changes in species composition – these are discussed later.

The relatively low numbers of small mammal species present in a given region make this group less suitable for investigations of the patterns of change in species richness as a result of fire than other groups, such as birds, that have higher species densities. Prodon *et al.* (1987) found a marked reduction in bird species richness in burned cork-oak forest in the Mediterranean, which quickly returned to near the levels of unburned control plots – within 6 months (Fig. 7.5). We might expect much slower changes in diversity of birds or mammals after fires in other vegetation types, such as those that take much longer to recover their diversity and structure.

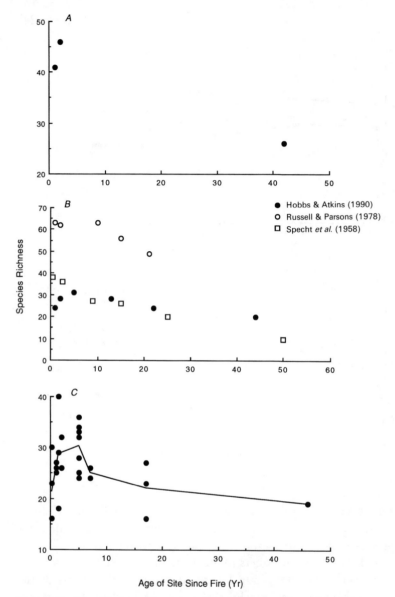

Figure 7.2 Changes in species richness over time since fire inferred from comparisons, at one time, of sites of different post-fire age (synchronic studies).
A. Californian chaparral sites (Vogl and Schorr 1972).
B. Three Australian sites in Western Australian *Banksia* woodland (Hobbs and Atkins 1990), eastern Australian forest (Russell and Parsons 1978) and South Australian *Banksia*-dominated heathland (Specht *et al.* 1958).
C. Replication of sites of various ages since last fire in Western Australian heathland – the line represents the mean species richness for each site age (Bell and Koch 1980).

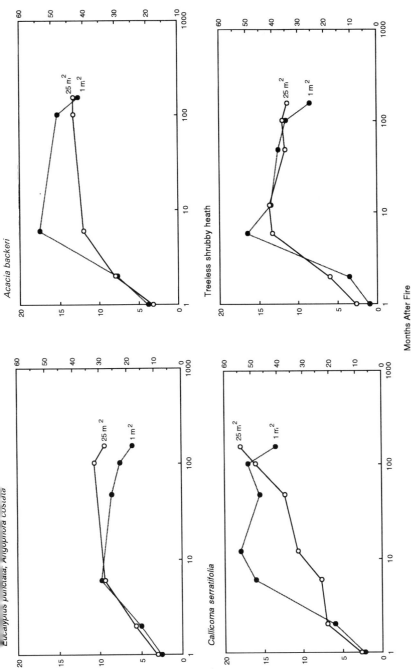

Figure 7.3 Relationship between the size of quadrat samples (1 m² versus 25 m²) and pattern of change in species richness after fire in four different plant assemblages in eastern Australian eucalypt forest (unpublished data from Muston 1987, with permission). The peak in species richness and decline that are evident in 1 m² samples are absent or much less pronounced at the 25 m² scale.

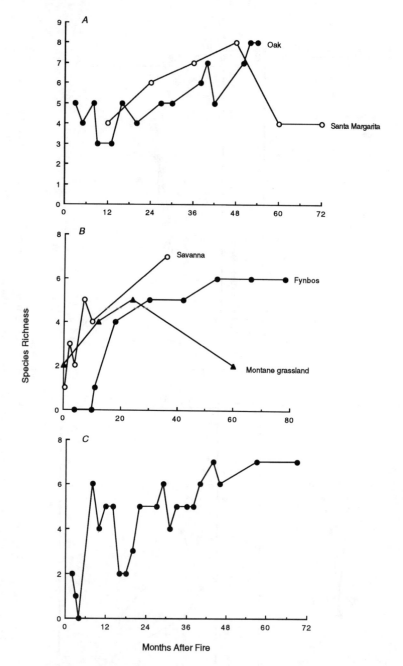

Figure 7.4   Changes in species richness of mammal communities with time after fire (after Fox *et al.* 1985).
A. Southern Californian chaparral (from Wirtz 1982).
B. Three African sites – savanna (from Kern 1978), fynbos (from Bigalke and Pepler 1979) and montane grassland (from Mentis and Rowe-Rowe 1979).
C. Eastern Australia (from Fox *et al.* 1985).

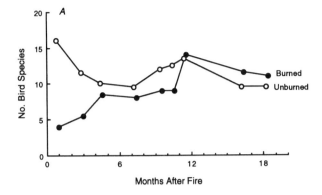

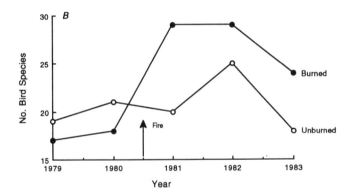

Figure 7.5   *A*. Recovery in 18 months after fire of bird species richness in a Mediterranean cork-oak forest (from Prodon *et al*. 1987). Data represent mean species counts from four line transects in a single burned and a single unburned site. After about 5 months following fire, species richness remained equivalent in the burned and the unburned site for the duration of the study (i.e. until June 1986).

B. In contrast, Christensen *et al*. (1985) found an increase in bird species richness as a result of fire in western Australian eucalypt forest.

Insects may offer even better opportunities for studies of animal community responses to fires. Studies on insects and other invertebrates reveal an array of responses of species richness to fire, perhaps contributed to by sampling technique (see Chapter 5). In some cases, species richness and diversity is seen to decline after fire (e.g. Force 1981) and in others, fire seems to stimulate an increase in richness (e.g. Whelan *et al.* 1980). Despite the fact that invertebrates offer great opportunities for well-replicated, manipulative studies of the effects of fire on the structure of animal communities, it is apparent that few studies have yet overcome the problems of high within-site variability (Campbell and Tanton 1981) and few have used adequate replication of the fire treatment (Tap 1994).

Changes in plant species composition

A change in species richness after a fire implies that there has been some change in species composition: some species may be lost and others gained. Is there a predictable pattern to changes in composition? Is invasion of a group of plant species with particular characteristics responsible for observed increases in species richness after fire? Is the loss of a suite of species with particular characteristics responsible for the declines in species richness often observed in sites that are left unburned for long periods? Of course, changes in community composition may also be occurring without any change in species richness, if species being lost are replaced, as they disappear, by new species. Trabaud and Lepart (1980) used the term 'fugacity' to describe the degree to which species composition changes over time, and developed an index of fugacity to quantify species turnover.

The most obvious pattern is the occurrence of a suite of annual and short-lived perennial species that temporarily occupy a site after burning, in some regions (see Muller *et al.* 1968, Specht *et al.* 1958, Kruger 1977, Naveh 1974, Trabaud and Lepart 1980, Trabaud 1987). These species may be well-dispersed, weedy opportunists that quickly occupy any disturbed sites (Trabaud 1987), and this view has been expressed many times as an explanation for the development of this annual/ biennial plant association as a result of fire. For example, Jarratt and Petrie (1929) stated that seeds germinating from the surface layers of ash in eastern Australian eucalypt forest 'were regarded as migrants from unburnt adjacent areas', and Ashton (1981) listed a range of herbaceous species, mostly characterized by having good long-distance seed dispersal, which can be found soon after fire in the tall eucalypt forests of south-eastern and south-western Australia. In the first year after fire in the Mediterranean maquis, vegetation is dominated by annuals that are opportunistic invaders not requiring fire to stimulate germination (Cody and Mooney 1978), and Granger (1984) reported that *Phillippia evansii* (Ericaceae) in South African forests appeared in recently burnt sites because of widespread dispersal of its fine seeds by the summer bergwinds.

As an aside, it is worth exploring whether the occurrence of an association of species dispersing into a site after fire is a general phenomenon (Whelan 1986, Trabaud 1987). Seed characteristics that would allow long-distance dispersal, small, light seeds with structures such as a pappus (Burrows 1986), are certainly prominent among pioneers in newly burned sites. The application of the common name 'fireweed' to well-dispersed, weedy plants such as *Epilobium* species is indicative of their early appearance and abundance in burned areas. Wildfire in the drier sites of Alaskan Taiga results in rapid dominance of a site by light-seeded species such as *Epilobium angustifolium* and *Salix* (Viereck 1973, Rowe and Scotter 1973). *Taraxacum, Hieracium* and *Chamaenerion* are the first taxa to appear after fire in Finland (Viro 1974); aspen and birch are wind-dispersed pioneer trees in the north-eastern USA (Little 1974); burrow-weed (*Aplopappus*) and snakeweed (*Gutierre-zia*) are post-fire pioneer shrubs that are disseminated by wind in the Sonoran Desert.

Despite these examples of apparent long-distance dispersal of seeds into recently burnt sites, few studies have demonstrated that such a process is common or widespread. In fact, a post-fire stage comprised mostly of well-dispersed, fugitive species appears to be lacking in many fire-prone regions. Kruger (1983), for example, summarized data from other sources that indicated that 'immigrant post-fire annuals' are not a conspicuous component of the post-fire vegetation in most Mediterra-nean-type vegetation. What is more, 'fire-ephemerals' were lacking in all studies surveyed by him except those in Californian chaparral. A flush of post-fire annuals also appears to be absent in vegetation of similar structure, the Florida sand-hill communities (Myers 1990). Hopkins and Griffin (1984) argued that whole families of plants that are characterized by having wind-dispersed seeds (i.e. Asteraceae, Poaceae and Orchida-ceae) are very poorly represented in Western Australian sand-heaths, a vegetation type that is certainly prone to fire.

In the chaparral, where a post-fire flush of annuals and biennials does occur, it seems that these species may result predominantly from dormant seeds stored in the soil (Sweeney 1956, 1968, Hanes 1977, Keeley 1981). Trabaud (1987) argued that two woody genera in Mediterranean ecosystems, *Cistus* and *Pinus*, have been viewed as typical well-dispersed pioneer species, which readily disperse into burned sites. However, his own studies (Trabaud 1980, Trabaud and Lepart 1980, Trabaud *et al.* 1985) suggest that seedlings of these species arise from seed banks stored within the burned site, or from seeds dispersed only a short distance from nearby adult plants that escaped burning.

The above discussion implies that, where a distinct post-fire plant association occurs, in many cases it arises from the burned site itself. This could occur by germination of dormant, long-lived seeds that are stimulated to germinate by the heat of the fire or other cues (Muller *et al.* 1968, Christensen and Muller 1975, Keeley and Keeley 1982), or by the temporary appearance above the ground of species that have long-lived, dormant tubers. Geophytes, such as some orchids, grasses and sedges, provide good examples of this, and a study by Muston (1987) in eastern Australian eucalypt forest showed that 17 of the 38 species that appeared after an experimental fire but were not recorded beforehand were geophytes. They must have been present prior to the fire – dormant and invisible. These data suggest that many species that appear to be absent from long-unburned vegetation but spring up immediately after fire may have been present all along as dormant roots or tubers. Seed dormancy has been much more thoroughly examined.

Long-term changes in the species composition of vegetation have been described in many systems. Species losses could occur through two main processes: (i) 'senescence' of old individuals of dominant species, without their replacement by invasion of other species (e.g. dominant species like *Calluna* in Scotland lose their ability to sprout with age (Hobbs and Gimingham 1987), and *Banksia ornata* and *B. marginata* in south-eastern Australian heathland senesce and degenerate without being replaced (Specht *et al.* 1958)); or (ii) through active replacement by species that invade slowly but are effective competitors.

Long-term changes in species composition through slow invasion of poorly dispersed species that were eliminated by a previous fire have been described in various studies. This process dominates descriptions of forestry in the United States (see Wright and Bailey 1982). The enormous complexity of effects of fire is illustrated particularly well in the Douglas fir chapter of Wright and Bailey's book. It is clear that fire intensity, forest species composition and location combine, perhaps with other factors such as insect attack, to influence the pattern of change in

Table 7.2. *Relative susceptibility to fires of conifer species in the Pacific North-west of North America*

Level of resistence	Common name	Species
Most	Western larch	*Larix occidentalis*
High	Ponderosa pine	*Pinus ponderosa*
	Douglas fir	*Pseudotsuga menziesii*
Medium	Grand fir	*Abies grandis*
	Lodgepole pine	*Pinus contorta*
	Western white pine	*Pinus monticola*
Low	Western red cedar	*Thuja plicata*
	Western hemlock	*Tsuga heterophylla*
	Noble fir	*Abies procera*
Very low	Subalpine fir	*Abies lasiocarpa*
	Pacific silver fir	*Abies amabilis*

Source: Modified from Starker (1934) and Wright and Bailey (1982).

species composition caused by fire. A detailed treatment of the effects of fire in these forests is outside the scope of this book, but this is covered in Wright and Bailey (1982), Wein and MacLean (1983) and many of the *Tall Timbers Fire Ecology Conference Proceedings*. Douglas fir forms associations with many other tree species, and Wright and Bailey (1982) explained the various patterns of post-fire vegetation change using the classification of fire responses proposed by Rowe (1983), namely:

Invaders   Well-dispersed weedy species with short-lived seeds
Evaders   Species with long-lived propagules stored in soil or canopy
Avoiders   Shade-tolerant species with slow reinvasion
Resisters Adults can withstand low-intensity fires, otherwise intolerant
Endurers   Sprouting species

The resistence of conifers in the Pacific North-west region to low-intensity fires varies markedly among species (Table 7.2, Fig. 7.6). Consequently, fires of different intensities may eliminate different subsets of these tree species. A good example of long-term changes in forest composition was described by Munger (1940) (see also Wright and Bailey 1982), for the Douglas fir/western hemlock forests of the western USA. Severe fire in a forest dominated by Douglas fir produces

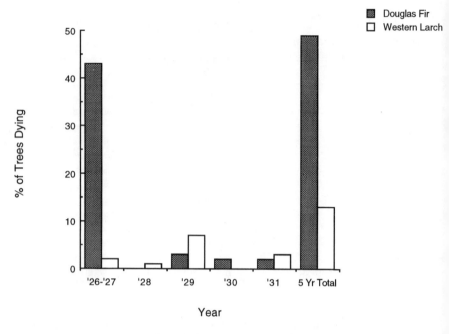

Figure 7.6 Varying mortality of tree species in a wildfire in North American conifer forest (from Starker 1934, after Wright and Bailey 1982), showing the relative tolerance of western larch and Douglas fir.

conditions suitable for Douglas-fir recruitment, and a dense stand can develop. The intense fire eliminates many other, less-tolerant tree species (see Table 7.2), but the thick bark of Douglas fir allows some trees to survive in patches where the fire was less intense. This, and the fact that many seeds are sufficiently well protected in cones, allow this species to attain dominance. Over the ensuing century, fire-sensitive but shade-tolerant tree species, such as western hemlock and western red cedar, invade and develop into a conspicuous understorey, eventually attaining dominance and virtually replacing Douglas fir in 500–800 years, barring interference by further fires.

Extensive fires in boreal forest areas of Alaska cause massive changes in the plant community, initiating what might be viewed as a classical secondary succession, on some sites, with well-dispersed species such as willows (*Salix* spp.) soon becoming established after the fire, species such as aspen (*Populus tremuloides*) and paper birch (*Betula papyrifera*) following, and white spruce (*Picea glauca*) stands eventually becoming re-established (Lutz 1956, Viereck 1973). Fires generally appear to be frequent enough to maintain large areas dominated by spruce, but long time periods without fire can lead to a marked change in the community, with sphagnum moss invading and altering the permafrost conditions so as to prevent return to previous stages.

Less-familiar examples of long-term changes in species composition come from fire-prone areas in which fire is not generally considered to be an initiator of typical successional processes: for example, the sandhill pine and scrub communities in the south-eastern USA (Myers and Ewel 1990); south-eastern Australian eucalypt forests (Gilbert 1959, Bowman and Jackson 1981, Ashton 1981), tropical monsoon-forests (Rose-Innes 1972, Clayton-Green and Beard 1985), the Californian chaparral (Keeley 1992), and even the *Calluna* heaths in Scotland (Gimingham *et al.* 1981).

The pine forests growing on sand in the south-eastern USA are typically intermixed with low-lying areas supporting hardwood vegetation (Fig. 7.7). In the long absence of fire, hardwood plants in genera such as *Quercus*, *Vaccinium*, *Myrica*, *Magnolia* and *Persea* invade a site and replace the formerly dominant pines (*Pinus palustris*, or *P. clausa*) (Laessle 1942, Veno 1976, Myers 1985, 1990). Myers (1985) made use of detailed vegetation maps made in 1932 by A. Blair at the Archbold Biological Station in south-central Florida, to quantify vegetation changes that have occurred over the past half-century, during which all fires had been prevented. Vegetation development on this site saw a variety of hardwoods invading and increasing in relative frequency, whereas originally only two pine and two oak species dominated (Figs. 7.8 and 7.9). This pattern of vegetation change, with an 'advancing front' of hardwood species moving into the pine-dominated sandhill and 'scrub' communities is now well accepted (Myers 1990, Abrahamson and Hartnett 1990, Platt and Schwartz 1990). A similar pattern of invasion of pineland by fire-sensitive hardwoods over long time periods apparently also occurs in the 'rocklands' environment of the south Florida Everglades (Snyder *et al.* 1990).

This scenario is remarkably similar a hemisphere away, in a completely different forest type and with a very different suite of plant species. Gilbert (1959) described the successional processes that occur over long time periods after fire in Tasmania (southern Australia). Like the Florida examples discussed above, the topography in Tasmania, and in many parts of the eastern Australian seaboard, supports a mixture of environments and associated plant communities. There has been some debate as to whether the mixture of distinct plant associations in these environ-

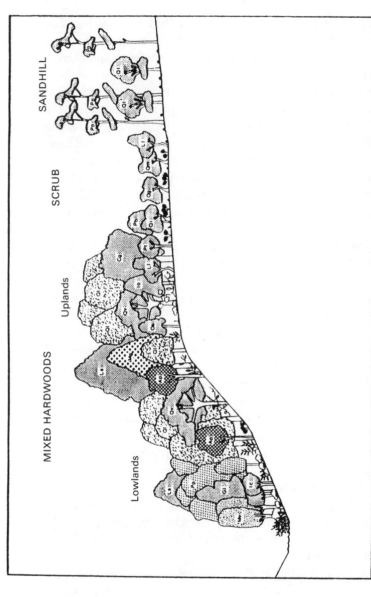

Figure 7.7 Schematic diagram of the topographic gradient and associated vegetation in central Florida longleaf pine sandhill and flatwoods communities (from Myers and Ewel 1990). This shows the dominance of hardwoods in the lowlands 'refugia' from which they are able to encroach on the pine-dominated scrub and sandhills in the absence of fire. AP = *Agarista populifolia*, Gl = *Gordonia lasianthus*, Ic = *Ilex cassine*, Io = *Ilex opaca*, Lf = *Lyonia ferruginea*, Ls = *Liquidambar styraciflua*, Oa = *Osmanthus americana*, Pb = *Persea borbonia*,

Figure 7.8 Photographs taken in 1929 (A) and 1988 (B) of a sandhill site that was originally pine and turkey oak and has seen the invasion of hardwood species over the 59 years. Reproduced, with permission, from Myers (1990).

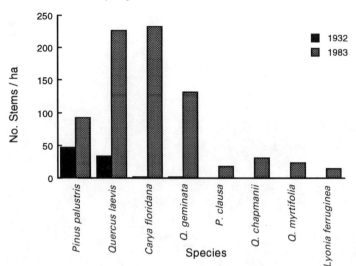

Figure 7.9 Change in species composition of the sandhill site depicted in Fig. 7.8. Tree density has increased overall, and hardwoods such as *Quercus* species have increased in relative abundance.

ments in Tasmania represent a static mosaic (see Jackson 1968, Mount 1979, Bowman and Jackson 1981). Nevertheless, it is apparent that mesophytic rainforest species can invade eucalypt forest that remains unburned for many decades (Ashton 1981). Dispersal of rainforest seeds by birds may be particularly important for these plant invasions (Gleadow and Ashton 1981, Whelan 1986).

California chaparral is a North American vegetation type which appears to replace itself following fires, and this is so clear that post-fire changes in vegetation have been referred to as 'autosuccessional' (Hanes 1971). A recent study by Keeley (1992) suggests that even very old chaparral (>200 yr since last fire) stands in which there has been little recent recruitment show little evidence of successional vegetation changes. Nevertheless, Lloret and Zedler (1991) found that the bird-dispersed evergreen shrub, *Rhus integrifolia*, a sprouter species that does not produce seedlings in immediate response to fire, recruits seedlings progressively between fires. In long-unburned sites, it may therefore contribute to a substantial shift in chaparral community structure.

In tropical northern Australia, eucalypt-dominated woodland (savanna) and monsoonal rainforests form a complex mosaic. It has been argued that the monsoonal rainforests were once more extensive, but that increased fire frequencies in the last 40,000 years have favoured the sclerophyll vegetation (Stocker and Mott 1981). In the absence of fire, some of the closed-forest species are expected to recruit into the savanna, causing an expansion of the boundary of the monsoonal rainforest patch (Woinarski 1990). Twelve years of experimental study at one site in northern Australia by Bowman *et al.* (1988), however, suggested that this sort of successional change in the absence of fire is not necessarily a general phenomenon.

It is worth looking for some general processes in these examples of vegetation changes in the absence of fire, in order to direct our thinking about other examples of vegetation change, such as woodland replacement of grassland in Africa (Walker 1982, Kruger 1983) and North America (Daubenmire 1968). The essential elements of this sort of vegetation change appear to be the following.

(i) A landscape that supports fire-prone vegetation that has been burned periodically, comprising species of plants that tolerate fire by sprouting and/or germination from a dormant seed bank.

(ii) An intermixing of other vegetation types, in topographic or other refuges from fires, in sufficiently close proximity to permit seed dispersal into surrounding vegetation. These species are susceptible to fire and can establish in the absence of fire, producing shade-tolerant recruits.

(iii) Mechanisms for dispersal of the more mesophytic species into the surrounding vegetation. The importance of animals in this process has apparently received little attention (but see Kauffman and Uhl 1990 for a discussion of the importance of frugivores in recovery of burned rainforest sites).

Gilbert (1959) pointed out one important part of this description of vegetation change in Tasmania that is relevant to all these case studies. Just as mesophytic plants remain in an otherwise fire-prone landscape in refugia that have particular topography, microclimate and soil fertility, so the whole extent of land dominated by fire-prone species such as eucalypts or pines is not necessarily available to the rainforest or hardwood species. Low soil fertility, in particular, is likely to provide a refuge for these species, even during very long fire-free periods. Bowman *et al.* (1988) argued that, if a high-frequency fire regime had persisted for long enough, the closed forest and sclerophyll forest species were likely to become specialized to particular site characteristics, such as edaphic conditions. After this sort of adaptive process, vegetation shifts following the removal or imposition of fire may not be expected to occur. Vegetation boundaries could therefore be 'the ghosts of fire regimes past' (apologies to Connell 1980).

Changes in animal species composition

There are a number of difficulties in evaluating the literature on change in wildlife species after fire. Forest fires and subsequent plant succession may vary. Hence, two burns may produce conflicting results as to the effect of burning. Methods of study also add variability. A count of species may not represent the effect of a burn because census areas and number of animals caught or observed are too small, studies are not long enough to detect a response, and transient species are called residents in burned or unburned habitats. Finally, different species of wildlife may be present in a recent burn than in an old one. *(J. F. Bendell (1974))*

Bendell's review in 1974 is comprehensive and insightful, and is worthy of close examination. He provided a good summary, which still applies, of the limits to our knowledge of the effects of fires on animal communities, emphasizing that species composition is very likely to change as a result of fire, and to continue changing for some time afterwards. The generality of a change in composition is clearly illustrated by his assessment of results of several studies of birds and mammals (Fig. 7.10), concluding that the post-fire mammal and bird communities lack some species that were present before burning and include a sometimes substantial number of species that were not present previously.

The explanation for the appearance soon after fire of a suite of species that were not present previously may be that there are groups of opportunist, generalist vertebrate species that flourish in recently burned areas. The contribution that the house mouse (*Mus musculus*) makes to the post-fire small-mammal community is probably in this category. Alternatively, these may be species that are habitat specialists, which now find the post-fire vegetation suitable, in terms particularly of food or cover. Bock *et al.* (1976), for example, found changes in the bird community of Arizona grassland and oak-savanna following both early spring and early summer fires (Fig. 7.11). The fires, which caused an increase in herbaceous plants and therefore of seed availability, favoured bird species that are grassland specialists, feeding predominantly on seeds. Shrubland species, such as the grasshopper sparrow, which appear to depend on shrubs for singing perches, were disadvantaged by fire (see Fig. 7.11). Similar patterns have been reported for other systems, such as

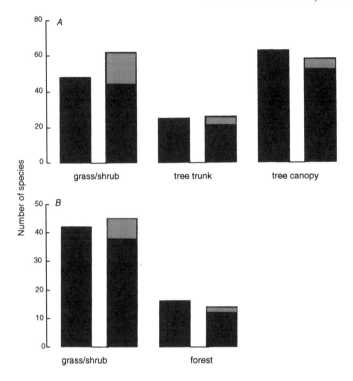

Figure 7.10 Changes in species richness of birds (*A*), and mammals (*B*) as a result of fire, indicating that the post-fire community (right-hand bar in each pair) comprises many, but not all, of the species that were present previously (dark portion of bar), and new species (stippled). The nature of the change varies among habitats, with grassland habitats gaining more species (based on data in Table V of Bendell 1974).

Californian chaparral/shrublands (Wirtz 1979), where species typical of grasslands temporarily move into burned chaparral, making use of the increased abundance of seeds resulting from the flush of ephemeral herbs. In the Mediterranean, Prodon *et al.* (1987) found a pattern of recoloniza-tion by birds of burned cork-oak forest that started with species typical of open vegetation (e.g. grasslands) being replaced by maquis species, and eventually by the forest-dwelling species that were present prior to the fire. This pattern is elegantly illustrated (Fig. 7.12) by relating the frequency of occurrence of particular bird species both to habitat type (from grassland, through maquis, to forest) and to time since fire. The wheatear (*Oenanthe hispanica*) clearly expands from grassland into

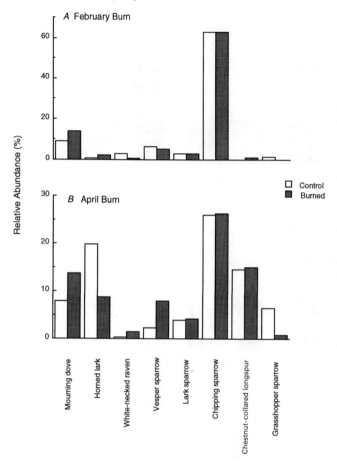

Figure 7.11 Changes in relative abundance of bird species in south-western USA grassland/shrubland communities burned in February (*A*) or in April (*B*). Seed-eating species such as the mourning dove were favoured by the fires while species depending on shrubs (e.g. the grasshopper sparrow) were disadvantaged. Data from Bock *et al.* (1976).

maquis after fire and its range contracts back to grassland about 5 years after the fire.

One important conclusion to come from several studies of bird and mammal communties disturbed by fire is that species richness may not change, and species may be neither lost nor gained. The major changes that do result from fire are in relative abundances. Recher *et al.* (1985), for example, found that a burned forest site in south-eastern Australian

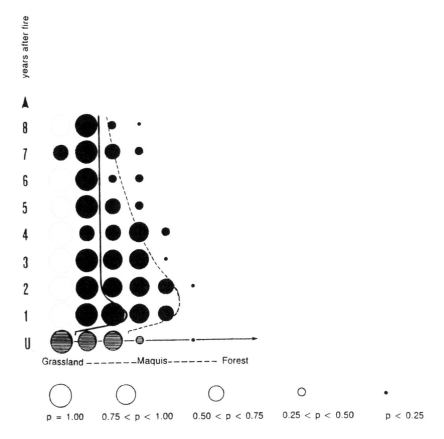

Figure 7.12 Change in frequency of occurrence (probability of encounter during a 20 min. point count) of a grassland bird species (*Oenanthe hispanica*) over time since fire in Mediterranean cork-oak forest (from Prodon *et al.* 1987). The probability of encounter is represented by the diameter of circles ranging from 1.0 (largest circles) to <0.25 (small points). *Oenanthe* occupies maquis (shrubland) sites soon after fire (horizontal axis), disappearing again over the years (vertical axis) as the shrubland recovers. The hatched circles represent the unburned situation. The *x*-axis represents the sequence type (grassland to forest).

eucalypt forests had lower population densities of birds, but only a slightly lower species richness, than an unburned site. However, relative abundances of some species decreased and others increased. Species with lower abundances in the burned site were typically ground- and foliage-foraging species such as thornbills (*Acanthiza* spp.), treecreeper (*Climacteris leucophaea*) and fantail (*Rhipidura fuliginosa*), while nectar- and lerpfeeders such as honeyeater and lorikeets were more abundant in the burned forest samples.

What might be the cause of changes in animal species richness and community composition after a fire? Since vegetation contributes so much to animal habitat, it is a good bet that vegetation changes will be reflected in changes to groups of animals. Hence, factors such as plant biomass, vertical structure, cover, and plant community composition are all likely to be influential in determining presence or absence of particular species of animals.

Changes in other factors: biomass, cover, vertical structure
Species diversity and species composition are but two measures of community structure. Biomass is another indicator that has been frequently measured. Although biomass cannot in itself yield much information about the ecological effects of fires, it is nevertheless important. Accumulation of biomass may indicate the rate of recovery of vegetation after a particular fire, or series of fires, but it also contributes to the likelihood, and the intensity and rate-of-spread, of the next fire.

Rate of increase in biomass after a fire clearly varies greatly among plant communties, even of a particular vegetation type, such as heaths (Fig. 7.13) or forests (see Fig. 2.11). Reasons for differences are no doubt manifold, but one which has received most attention is the effect of a dominance of obligate-seeder species versus sprouters in the plant community (Bell 1985). Rapid recovery of biomass clearly occurs when most regeneration is sprouting from protected tubers, roots, stems and branches. Hence, grasslands rebuild biomass very rapidly. However, Schlesinger and Gill (1980) reported a very rapid increase in biomass after fire in Californian chaparral that was dominated by reseeding species.

In the light of discussions about the relative importance of invasion of species versus germination and sprouting from within a burned site, it would be interesting to compare biomass accumulation curves after different sorts of disturbances in fire-prone environments. One would predict (e.g. Pate *et al.* 1984) that sprouting and dormant seed banks would give fire-prone commmunities better opportunity to regain biomass after fire than after other sorts of disturbance.

As discussed in Chapter 2, the horizontal and vertical distribution of biomass, and the relative proportions of living and dead biomass, are important in determining the characteristics of the next fire, such as type

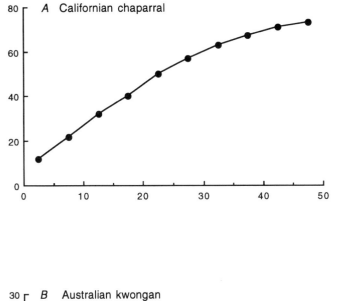

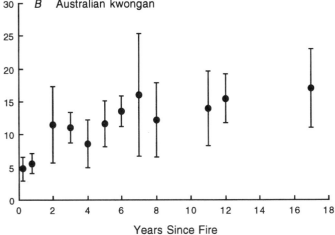

Figure 7.13 Accumulation of biomass in heaths, showing a steady increase over time since fire, perhaps approaching a steady state after many years. *A.* from data of Kessell and Cattelino (1978) for Californian chaparral. *B.* from data of Rullo (1982, after Bell *et al.* 1984) for Western Australian kwongan heathland.

of fire, fire intensity and rate of spread, yet these variables are rarely quantified.

Bell *et al.* (1984) reviewed various studies of post-fire increases in cover, and found a marked difference between Californian chaparral and Australian Mediterranean-climate heaths. The first years of chaparral community development after fire are dominated by herbaceous species, and it takes many years for even the sprouting shrub species to achieve substantial cover. This produces a bimodal pattern of cover over time since fire. Clearly there are two processes at work here: (i) the productivity of a site can determine the rate of accumulation of biomass and hence cover; and (ii) the species composition can determine the rate of increase in cover, with ephemeral herbs and sprouters both able to produce high levels of cover rapidly.

In the context of biotic interactions within a community, the importance of the build-up of biomass (and hence other characteristics of vegetation structure such as cover) is perhaps its influence on animals living in the community. Fox and Fox (1987) summarized their work, with various colleagues, on the responses of the small-mammal community to fire in eastern Australian forests and shrublands (Fox and McKay 1981, Fox 1982, 1983, Fox *et al.* 1985, Fox and Fox 1987). These studies emphasize the correlation between the development of post-fire vegetation structure and the reappearance of small mammal species. *Pseudomys novaehollandiae* (a native rodent) and *Mus musculus* (introduced) dominated a forest site immediately after fire, at a time when cover was low at all vegetative levels (Fig. 7.14). As biomass and percentage cover increased, various species reached their peak abundance.

Fox and McKay (1981) examined correlates of increases in total biomass of one native rodent, *Rattus fuscipes*, in their study of changes in small-mammal community structure after fire in south-eastern Australian heathland/woodland. The major part of the variance in the biomass of *Rattus* was explained by variation in the amount of leaf litter present. Other studies have shown different associations for other species; for example, *Pseudomys novaehollandiaea* was associated with the presence of an understorey layer represented by a high diversity of heath shrubs (Fox and Fox 1978). Using approaches such as these, it should be possible to build up a multivariate picture of the variation in post-fire vegetation that best explains the variation in mammal species composition.

Based on the above discussion of differences, between chaparral and Australian forests, in the rates of accumulation of cover and biomass, a

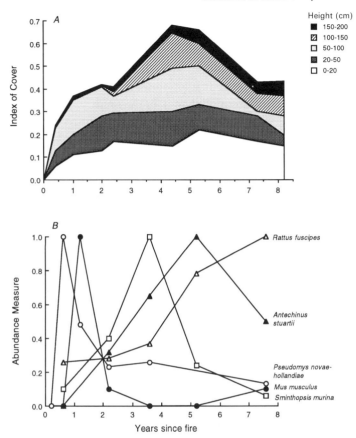

Figure 7.14 In eastern Australian heathland vegetation, patterns of change in the mammal community may be associated with changes in cover at various vegetation levels (data from Fox and Fox 1987).
A. Cover provided by lower layers of vegetation recovers quickly after fire, while cover of layers above 1 m starts to increase only after about 3 years.
B. The return of some mammal species after fire (e.g. *Antechinus* and *Sminthopsis*) appears to be associated with the return of cover at the higher layer while other species (e.g. *Mus* and *Pseudomys*) can invade earlier.

different pattern of development of the mammal community after fires might be expected. Fox *et al.* (1985) compared the results of several studies on mammal communities in Californian, South African and south-eastern Australian shrublands. It was not possible to detect differences in the patterns of recovery of the mammal community: changes in mammal species richness varied among studies within a region far more than the general differences between continents. This is not surprising, recognizing that there are bound to be large differences in techniques of study, intensity and duration of trapping, areas surveyed, etc., and manipulative studies within a particular region would be needed in order to examine the role of post-fire plant community changes in controlling the pattern of change in the fauna.

Bendell (1974) made some interesting comments about the possible effects of animals in determining the fire regime of an area, reminding us that the interactions between fires, vegetation and animals are indeed complex. He argued that feeding by squirrels in conifer forests could cause alteration to particular trees, thus increasing the likelihood of lightning strike. Moreover, piles of cone scales at the bases of certain trees may enhance the spread of fire from a lightning strike and would certainly create great spatial variation in fire intensities, once a fire was spreading (see also Lutz 1956). Grazing by large herbivores can, of course, maintain the standing crop in grasslands at a low level, hence keeping fuel loads low and reducing both the likelihood and intensity of fire. In a more complex argument about North American forests, Bendell (1974) considered the possibility that selective grazing by mammals, such as deer, on fire-sensitive, shade-tolerant tree species (e.g. hemlock, cedar and balsam fir) helped maintain a forest in a flammable state (e.g. dominated by pines, black spruce), thus maximizing the possibility of future fires that would provide better quality 'green pick' in general and also enhance the growth of specific browse species. This is, of course, a difficult hypothesis to test, and it is not clear that deer would have the luxury to feed selectively on browse species that were not optimal in terms of nutrient content and digestibility, principally to increase the possibility of some future fire.

History
Several studies of plant succession in relation to fire have indicated that succession follows the initial floristic composition model of Egler (1954). The implication here is that the pattern of vegetation change after the fire is determined largely by the species composition immediately prior to the fire (or perhaps immediately after it; see Trabaud 1987). Thus, the history of a site might be expected to have a strong influence on the structure of the plant community that develops after the fire. One aspect of site history that is easily measured and might be important is the time since the last fire.

Hobbs and Gimingham (1987) showed that the time since the

previous fire was a major factor influencing the development of the post-fire community in Scottish heathlands. In these heathlands, species diversity generally increases as an immediate result of fire and then declines as *Calluna* regains dominance and opportunist species disappear. When vigorous *Calluna* plants are burned, sprouting allows this species to regain dominance rapidly. However, about 15 years after fire *Calluna* plants lose the ability to sprout when burned (Mohammed and Gimingham 1970). Fire intensity is typically high in older *Calluna* stands (Hobbs and Gimingham 1984), perhaps contributing to the poor sprouting (see Chapter 3). Whatever the reason for poor sprouting, recovery of this species after fire in older stands occurs solely through seeds, leaving opportunities for other species to occupy the site. Hobbs and Gimingham (1984) showed that when *Calluna* regeneration is very slow, for example after a fire in a 'degenerate' stand, species such as *Vaccinium myrtillus* may attain dominance and thence inhibit recovery of *Calluna* by preventing establishment of *Calluna* seedlings (Fig. 7.15). In other sites, bracken fern (*Pteridium aquilinum*) may also invade after an old *Calluna* stand has been burned, impeding recovery of dominance by this heath species (Hobbs and Gimingham 1987).

Herb and grass species in these Scottish heathlands may respond to the increasing dominance of a burned site by *Calluna*. As plant densities decline, there will be increasing reliance on soil-stored seed for recovery after a fire. Seeds may remain viable only for a short time (Mallik 1984b), so a recovery of these species after fire in an old *Calluna* stand may have to be via seeds dispersed into the site. As this may take some time, and as the *Calluna* canopy will soon reform, greater post-fire species richness after fires in young *Calluna* stands than in degenerate stands would be predicted. This is exactly what was found by Hobbs *et al.* (1984) (Fig. 7.16), when species richness in 1 m^2 quadrats was compared in the year after fire for a set of sites of differing ages. Of course, these data must be interpreted carefully, in the light of the comments made above about 1 m^2 quadrats measuring species *density* as much as species richness of a community.

The above illustrations of the importance of fire history in determining the immediate post-fire community point to the need for experiments to determine the effects of various fire regimes on communities. There is potential for enormous variation in 'prior condition' of the community, such that it may even be the most important determinant of the post-fire patterns of community change. Replication of experimental studies with similar fires in sites with different histories, and

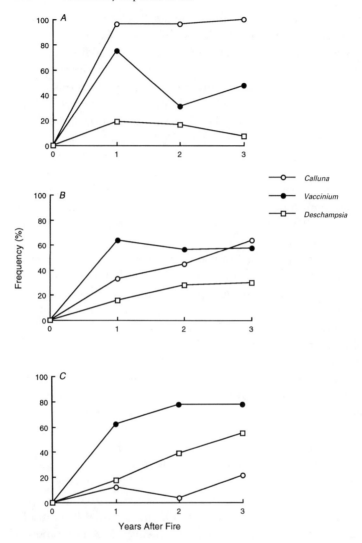

Figure 7.15 Recovery of several co-dominant species in Scottish heathland after fires in three stands that were 6 years old (*A*), 10 years old (*B*), and 15 years old (*C*) after a previous fire. *Calluna* recovers dominance rapidly in sites previously burned only a few years ago (*A*), while *Vaccinium* can gain dominance if a site is left unburned for 15 years (*C*). From Hobbs and Gimingham (1987).

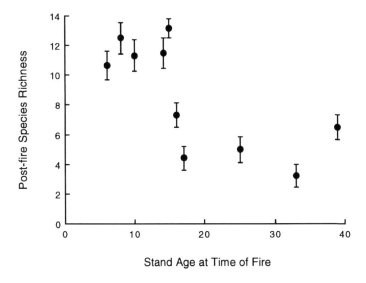

Figure 7.16 The influence of stand age at the time of a fire and the species richness of vascular plants one growing season after fire. Data are mean number of species (±SE) in four 1 m² quadrats in each stand. From Hobbs and Gimingham (1987), after Hobbs *et al.* (1984).

replication of fires over several years in sites with similar histories is critical to dissecting out the importance of history in post-fire community change.

Variation in fire regime

Some studies have attempted to manipulate a component of fire regime and have then measured some relevant community parameters. As was pointed out in Chapter 4, this should be conducted with replication of treatments. Two main problems are: (i) that the ideal experiment, with replication and manipulation of various components of fire regime, is a very large one; and (ii) that unexpected fires all too frequently interfere with the best laid plans.

Problems with the large scale of the experiments that are needed have various parts. First, few individual researchers have the security of tenure to contemplate setting up long-term experiments. Second, the magnitude of community-wide experiments now clearly requires a team of investigators, and few institutions are in a position to guarantee this sort of support both for the setting up of a long-term experiment and also for ongoing maintenance and monitoring. Third, in order to conduct a community-wide experiment, individual treatment plots are necessarily much smaller than areas usually burned in wildfires. Consequently, some outcomes in an experiment may be artifacts of the sizes of treatments. There are several good examples of fire experiments that illustrate some of these points. The following examples are by no means exhaustive, but they are selected to illustrate particular strengths and difficulties in these sorts of studies.

Fire experiments in African ecosystems

Ecological field experiments are likely to be conducted on an increasing scale in South Africa, and modern experimental designs and interpretations should be applied to such experiments in order that the maximum amount of reliable information may be extracted from the observational data.... The improved methods were first applied to agricultural field experiments and they are generally used in agricultural research in South Africa. Their advantages have, however, apparently not yet been fully appreciated by South African ecologists. (*Wicht (1948) p. 480*).

In 1945, Wicht (1948) set up a randomized block experiment to examine the effects of different seasons of burning in a South African sclerophyll shrub community. It is a little surprising that the introduction to that study, written by Wicht in the mid-1940s, should still be pertinent today:

Wicht set up eight burning treatments, with fires in the middle of each of the following months: January, February, March, April, September, October, November, December. Treatment quadrats were 2 m square, with a 2 m strip between plots, and treatments were randomly allocated to the eight quadrats in each of the five blocks. He also discussed the advantages and disadvantages of the size of treatment quadrats, the main advantage being that a reasonable number of replicate quadrats for each treatment could be accommodated within what was perceived to be a homogeneous tract of vegetation. The main disadvantages were suggested by Le Maitre (1987) to be the non-representative nature of the fire intensity that could be generated in such small plots (fuel had to be added to some quadrats to ensure a 'clean' burn) and the fact that only a single large overstorey shrub (most commonly *Protea burchelli*), and only scattered representatives of other important fynbos species, were present.

Wicht (1948) provided some preliminary analysis of the results of this experiment, examining the effects of the summer and autumn fires

(January to April). He clearly planned to continue the study: 'It is proposed to publish further accounts of experimental results from time to time.' This did not eventuate, until the recent analysis by Le Maitre (1987) of Wicht's data. This statistical analysis did not take advantage of the randomized block design set up by Wicht, but ignored the variation among replicate quadrats by presenting only totals or means. The results showed that season of burning did not have much of an effect on the species richness per 4 m² quadrat but did appear to affect the density of various species or groups of species (Fig. 7.17).

There seems to have been a tradition of setting up replicated fire experiments, based on small plots, in Africa. West (1965) summarized a large number of experimental studies commencing with the work of Phillips in South Africa in the early 1920s (Phillips 1919, 1920), and including a range of studies from western and eastern Africa. Perhaps the nature of savanna grasslands and relatively homogeneous shrublands makes these ecosystems good targets for experimental plant ecology, based on experiments in agronomy, in much the same way as marine intertidal experiments have come to dominate more recent ecological studies in community ecology.

In summarizing the grassland studies, West (1965) emphasized the effects of protection from fire. Changes in species composition typically occur, with some species, for example *Themeda triandra*, disappearing very rapidly when fire is excluded. Season of burning appears to be important in determining the direction of the changes in plant species composition that occur. Fires in the dormant season maintain *Themeda* and exclude herbaceous weeds and woody invaders. In contrast, fires in the growing season exclude grass species which are viewed as 'favourable', such as *Themeda*. These changes appear to be associated with the effects of the accumulation of undecomposed litter (see Chapter 5).

Australian wet–dry tropics: Kapalga and Munmarlary
The problem of small sizes of burned treatment quadrats has been recently addressed in a large-scale experiment set up by the CSIRO in northern Australia (Braithwaite 1990, Lonsdale and Braithwaite 1991). The Kapalga Research Station comprises 700 km² within the Kakadu National Park, and the southern half of Kapalga has been divided into 12 compartments that will be subjected to a range of fire treatments. Compartments are 15–20 km² and each is a separate water catchment. They provide the opportunity for a randomized block experimental design with four treatment compartments arranged in three blocks,

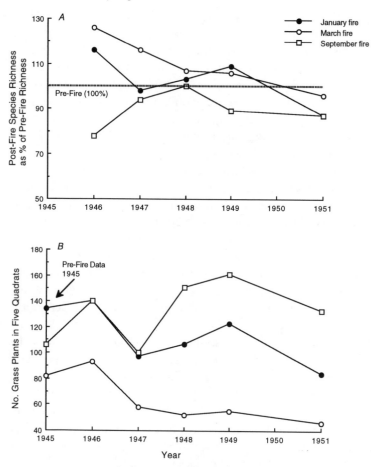

Figure 7.17 A. Change in species richness (mean of five 4 m² quadrats) expressed as a percentage of the mean pre-fire richness in these five quadrats (data from Le Maitre 1987), for three of the eight experimental fire seasons set up by Wicht (1948).
B. Change in numbers of individual grass plants in five 4 m² quadrats for Wicht's experimental fires (data from Le Maitre 1987).

hopefully overcoming the relative insensitivity of small, variable, sample plots in detecting subtle differences between treatments, unless very large numbers of replicates are used (Lonsdale and Braithwaite 1991).

There are four fire treatments: (i) annual early dry-season fires, typically of low intensity (May–June); (ii) annual late dry-season fires, typically of high intensity (September); (iii) annual progressive fires through the compartment, lit early in the dry season at first and followed up by other ignitions as vegetation dries out progressively downslope; (iv) no human-lit fires, which will produce a low-frequency, wet-season fire regime, whenever wet-season lightning storms manage to cause ignition in these compartments.

One objective of this team-based, long-term study is to obtain an understanding of the distribution, abundance and persistence of plant communities in the savanna woodlands of tropical Australia, integrated with studies of small vertebrates, the insect fauna, plant–herbivore interactions and predator–prey interactions.

The establishment of the Kapalga study follows a similar study in northern Australia, run by the Conservation Commission of the Northern Territory at Munmarlary (Bowman *et al.* 1988, Westoby 1991). Four different fire regimes (annual fires in early and late dry season, biennial fires in early dry season and no fires) have been applied to each of three replicate 1 ha plots. This study has revealed no substantial shift in floristics following fire exclusion, though vegetation structure has changed substantially. However, more recent studies at other sites (e.g. Bowman and Fensham 1991, Bowman 1993) suggest that the increase in woody shrubs surrounding patches of monsoonal rainforest that follows cessation of burning may facilitate establishment of rainforest species in the long term.

North Florida pinelands: the Tall Timbers experiment
The effects of frequency of burning was the focus of long-term demonstration plots set up at the Tall Timbers Research Station in northern Florida, USA. Begun in 1959, replicated 0.2 ha plots were removed from the previous management regime of annual winter burning and allocated to one of six fire-frequency treatments, with burning in March: intervals of one, two, three, seven and twenty years and no fire. These plots were set up in second-growth pine forest. It has been argued that frequency of fires, as well as the intensity and season, may control the nature of the forest community able to occupy the site.

After 23 years, Platt and Schwartz (1990) surveyed these plots (two plots per fire treatment), focussing on species composition, relative abundance and physiognomy. Twenty three years of annual burning reduced the percentage cover, and the species richness and diversity of woody species considerably, compared with the unburned plots (Fig. 7.18) and also confined most of the woody vegetation to a low layer of

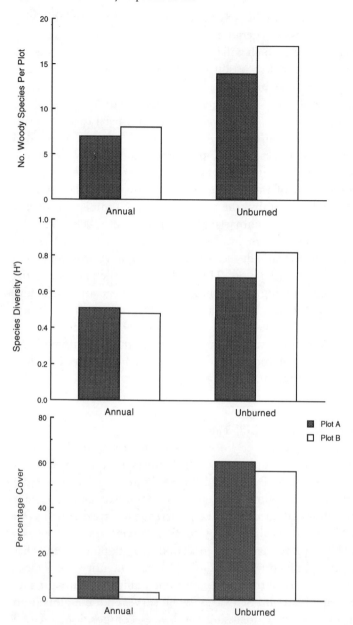

Figure 7.18 Comparison of the effects of annual burning versus no burning in the Tall Timbers pinelands experiment (from Platt and Schwartz 1990). Species richness, species diversity (H') and cover were all greater in both of the replicate unburned sites than in both of the annual-burning sites. Replicates described as 'Plot A' and 'Plot B'.

sprouts from root stocks. The less extreme fire frequencies did not, however, significantly influence percentage cover or species diversity (Fig. 7.19), suggesting that varying the fire interval within this range would have only subtle effects on the development of a hardwood-dominated forest.

As an aside, it is worth mentioning here a long-term study set up in 1946, in a 40-year-old loblolly pine plantation in South Carolina (the 'Santee fire plots'). This study is still continuing (Robbins and Myers 1992). Fire treatments are annual summer fires, annual winter fires, biennial summer fires, periodic summer fires and periodic winter fires. Although set up in a plantation forest, by the 1980s this study had shown clearly that season of fire strongly influenced community composition by regulating the growth of hardwoods. Annual summer burning had reduced stem density of hardwoods to less than 2000 per ha. Plots subjected to annual winter burning sustained a density of hardwood stems of over 45000 per ha. Though the study is experimental, long-term and shows striking results, none of the reports describing it has been in the mainstream ecological literature (Chaiken 1952, Lotti *et al.* 1960, Langdon 1981, Waldrop *et al.* 1987).

The Mediterranean Garrigue: Trabaud's experiments
Trabaud (1974) set up a long-term experiment near Montpellier in the south of France, designed to compare the effects of different seasons and frequency of fire on a Mediterranean Garrigue community, dominated by *Quercus coccifera*. In 1969, five replicate plots, 20 m × 5 m were established in each of six season × frequency combinations (see Trabaud and Lepart 1981):

Factor 1: season of burning
    spring fires
    autumn fires
Factor 2: frequency of fires
    every 2 years
    every 3 years
    every 6 years

This heathland community is dominated by a shrub-like oak species, *Quercus coccifera*, that sprouts readily after fire. Trabaud and Lepart (1981) found that none of the season × frequency combinations caused drastic changes in species richness over the first experimental cycle of 7 years (Fig. 7.20)

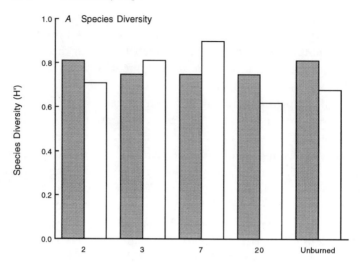

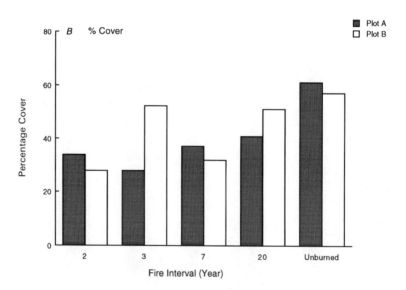

Figure 7.19 Response of species diversity – *H'* (*A*) and percentage cover (*B*) after 20 years of exposure to the range of experimental fire frequencies in the Tall Timbers experiment (data from Platt and Schwartz 1990). Species diversity did not differ markedly among these treatments, although percentage cover was slightly less at higher fire frequencies.

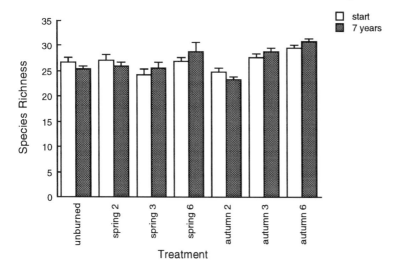

Figure 7.20 Changes in species richness in Mediterranean garrigue after 7 years of experimental fire frequency treatments in spring and autumn (data from Trabaud and Lepart 1981).

It may be argued that the simple measure of species richness hides a lot of information in this sort of study. A constant level of species richness, for example, may disguise a marked change in the composition of a community as secondary succession proceeds. Trabaud and Lepart (1981) therefore calculated a 'fugacity index', based on the species that were present at the start of the experiment (i.e. a reasonable time after the previous fire) but which disappeared during the 7 years of the study, and the species not present in a site at the start but which appeared thereafter. Each fire was followed by an increase in the fugacity index, caused by an influx of species that did not persist through subsequent samples. Lowest indices of fugacity (i.e. representing the sample times in which a plot was dominated by species with an expectation of long persistence) occurred immediately after fire, when all but the long-lived sprouters had been removed, and in the oldest post-fire samples of each experimental treatment, after short-lived species had disappeared. All experimental treatments except one, 2-yearly burning in autumn, showed stability in the cycle of fugacity, at least for the relatively short duration of the study. The 2-yearly autumn fires produced an increase in fugacity, both at each successive post-fire 'trough', and at each successive peak, indicating a relative decline in long-lived, persistent species, and a continuing input of new invaders between fires.

Other experimental studies

Even though undertaking an experimental study manipulating some aspect(s) of fire regime is clearly ambitious, there are several published examples, in addition to the above, that have received less attention. Clark (1988) described a 7-year study in eastern Australian tall open eucalypt forest. Four replicate 5m × 5m plots were allocated to three treatments: unburned; burned in spring; burned in autumn. After 6 years, the burned plots were re-burned, in accordance with the land management strategy for the area at the time, and the vegetation monitored for a further year.

One conclusion of Clark's study is that there is an enormous variation in responses to fire among species, between the two seasons of burning and between the two fires, 6 years apart, for each fire season. This variation emphasizes the importance of post-fire site conditions, such as rainfall, in determining the reaction of a population to fire (see Chapter 5). Nevertheless, Clark was able to classify species into the following groups according to their population responses to fires, compared with the changes in populations in the unburned sites:

(i) Decreased after both spring and autumn fires;
(ii) Decreased after autumn fires but not after spring fires;
(iii) Decreased after spring but not after autumn fires;
(iv) Remained stable after fires in either season;
(v) Did better after both sorts of fires.

The overall pattern was that recovery of both number of species and population sizes was better after the autumn than spring fires in the first experimental burns. However, this pattern was reversed after the second fires, 6 years later. Species that recovered better after the spring fires showed good survival of established plants as well as recruitment, whereas those species that recovered better after autumn fires did so solely because of better recruitment.

Several authors have focussed on both fire frequency and fire season as factors that might determine the relative proportions of obligate-seeder and sprouter species in a plant community. Baird (1977), for example, argued that, in south-western Australian coastal plain sites, sprouters are favoured by cool-season fires whereas obligate seeders will generally be favoured by summer and autumn fires.

The apparent relationship between fire frequency and relative success of sprouters and obligate seeders has led researchers in two continents, Africa and North America, to propose qualitatively similar models (Kruger and Bigalke 1984, Keeley and Zedler 1978, Keeley 1981, Le Maitre 1992). These models suggest that communities with low or high fire frequencies are likely to be dominated by sprouter shrubs while communities experiencing intermediate fire frequencies are more likely to be dominated by obligate seeders. In Australia also, Christensen and Kimber (1975) adopted this sort of model in arguing that south-western Australian Jarrah (*Eucalyptus marginata*) forest, sites which had low proportions of obligate seeders, have historically had higher fire frequencies than Karri forest (*E. diversicolor*), with a high proportion of obligate seeders (see also Christensen and Annels 1985).

Inference from comparisons of sites with different fire histories

Even without the opportunity for experimental studies, some useful information may be obtained from carefully designed and interpreted synchronic studies. One good example is a descriptive study by Nieuwenhuis (1987), which was designed to estimate the effects of different fire frequencies (high versus low) on the relative abundances of obligate seeder and sprouter species in a Hawkesbury Sandstone plant community near Sydney, Australia. This study took advantage of a large National Park with good fire-history records. Paired sites in close proximity were established in 15 locations. One of each pair had been burned several times (3–4) in the past 20 years, the other either once (in 1965) or twice (1965 and only once since). Overall, the comparison of percentage ground cover by obligate seeders and sprouters shows that the latter have been disfavoured by frequent burning (Fig. 7.21).

A study by M. D. Fox and B. J. Fox (1986) provided essentially similar results. Only two sites were compared here: both were burned in 1968, one had been subsequently burned only once – in 1980, the other had been burned twice – in 1974 and again in 1980. For obligate-seeder species, more than expected were found in the site burned only once; for sprouters, more than expected were found in the site that was burned twice (Table 7.3). Estimates of cover support the conclusions of Nieuwenhuis (1987) above, namely, that the cover of obligate-seeder species is less in sites that are burned more frequently (Fig. 7.22).

Although these conclusions are derived from a single study with no replication of burning treatments, and the observed differences between

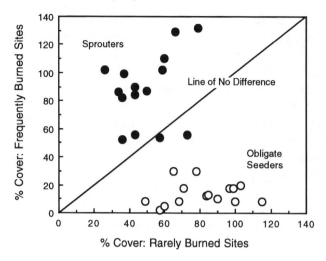

Figure 7.21 Comparison of relative abundance (percentage cover) of sprouter and obligate-seeder species in eucalypt woodland near Sydney, Australia, in a set of 15 matched pairs of plots. Each point represents the percentage cover in the plot burned frequently during the past two decades (*y*-axis), and the percentage cover of the counterpart plot burned just once or twice (*x*-axis) (from Nieuwenhuis 1987). Obligate seeders had greater cover in the rarely burned plots while seeders had greater cover in the frequently burned plots.

the two sites may be due to any number of unstudied factors, they do suggest that marked changes in community structure, for example in the relative frequencies of obligate-seeder and sprouter species, may follow from a single, short, inter-fire period. An extreme example of this was discussed in Chapter 2 (see Fig. 2.18; Zedler *et al.* 1983) for Californian chaparral.

Inferring the effects of fire regime by comparing sites of different fire histories has been the most common approach for studies of animal communities as well as for plants. Springett (1976), for example, examined the soil fauna of two Jarrah (*Eucalyptus marginata*) forest sites in south-western Australia. One site had not been burned for over 40 yr and the other had been subjected to 5-yearly hazard-reduction burning and was sampled in the 5th year, just before the next fire was due. The long-unburned site showed much higher densities of soil arthropods and species richness was 28, compared with 14 in the frequently burned site. These differences were attributed to the fact that litter mass and cover were maintained at low levels by the regular burning. The main limitation with this study is the lack of replication of frequently burned

Table 7.3. *Comparison of the numbers of species of obligate seeder and sprouter shrubs recorded in sites burned once in the previous 12 years, in those burned twice in this time, and in both sites*

	No. species present in site		
Regeneration strategy	Burned once	Burned twice	Both sites
Obligate seeders	5	2	11
Sprouters	0	9	11

Source: From M. D. Fox and B. J. Fox (1986).

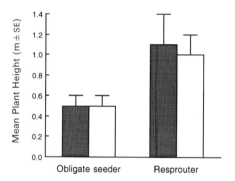

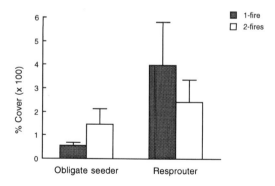

Figure 7.22   Comparison of mean plant height (*A*) and percentage cover (*B*) of obligate-seeder and sprouter species in two sites, one burned only once since 1968 (in 1980 – stippled bars) and the other burned twice in this time (1974 and 1980 – open bars). Sprouters had a greater cover in the sites burned only once, while the obligate seeders had greater cover in the site burned twice (data from M. D. Fox and B. J. Fox 1986).

and rarely burned sites, though Springett (1976) argued that studies of invertebrates have great potential in the monitoring of community changes as a result of fire, because samples can be obtained quickly and sorted sufficiently well to recognisable taxa that measures of community structure, such as species diversity, can be used.

Studying the bird community of Jarrah forest, Christensen *et al.* (1985) found that species richness was affected differently by a mild fire in spring and a high-intensity wildfire in summer. The differences in richness were slight, however, and difficult to interpret because there was no replication of fires and the sites were studied at different times after their respective fires. Of more importance is the fact that certain groups of birds responded differently to the different fires (Fig. 7.23). Densities of ground foragers such as the robins (*Eopsaltrica* spp. and *Petroica multicolor*) were reduced in both burned sites compared with a long-unburned site, though by much more in the wildfire site. Density of shrub and mid-storey foragers was greatest in the cool-burn site, and did not appear to be greatly influenced by the high-intensity fire. To complete the contrasts, densities of overstorey leaf gleaners showed the greater reduction in the cool-burn site.

Mechanisms of community change

If species are lost over time after fire, what processes cause the loss? If species accumulate, what processes determine the timing and rate of acquisition? What processes determine the relative abundances of species? It is tempting to view changes that occur after fire as an example of 'typical' community change after disturbance, and therefore to look to succession to explain the patterns. The literature on post-fire succession includes many apparent attempts to make observations fit expectations. In some regions, fire undoubtedly causes comprehensive removal of species and initiates a successional process of predictable, directional community development (e.g. northern boreal forests; tropical rainforests). In others, however, the mechanisms of post-fire vegetation change should be examined more thoroughly before the term succession is invoked to describe the observed changes.

Succession

The difficulties in interpreting changes in communities after fires lie not only in a desire to see fire as simply another type of disturbance initiating

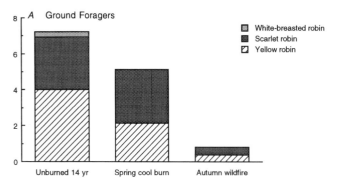

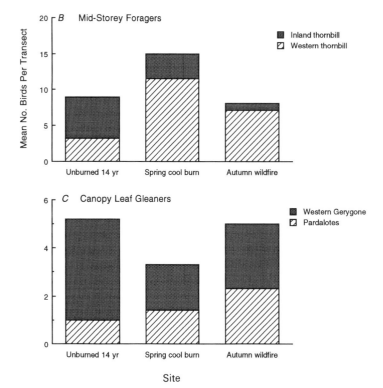

Figure 7.23 Mean densities per transect of bird species in three groupings: ground foragers (*A*), mid-storey foragers (*B*), canopy gleaners (*C*) (modified from Christensen *et al.* 1985). The effects of different types of fires varied among groups of species.

a classical successional process, but also in the lack of clarity surrounding the theories of succession. It is worth exploring these theories further. It is firstly important to distinguish between pattern and process in relation to vegetation change, both of which have been labelled 'succession'. The pattern is simply a description of the change in community composition over time since disturbance. The changes may be predictable, to some extent, but represent, in any case, a succession of species occupying a site. Measurements of abundances or percentage cover of each species will reveal a repeating sequence of appearance, increasing importance, decreasing importance and disappearance. Readily described patterns of plant community change may be found after most disturbances.

The recognition of a pattern of change should be separated from inference of the processes producing the change. Unfortunately, the neatness of the facilitation hypothesis (see Odum 1969, Connell and Slatyer 1977) or 'relay floristics' (Egler 1954) has been seductive, and the description of vegetation change after a disturbance as 'succession' carries, for many people, connotations of certain processes (Egler 1981). These processes include some or all of the following: (i) removal of many species from a site by the disturbance; (ii) different times of invasion of propagules; (iii) alteration of physical conditions by early invaders; (iv) dependence of later invaders on changed conditions before successful establishment of propagules; (v) competitive interactions between earlier and later invaders.

Recent reviews (e.g. Drury and Nisbett 1973, Connell and Slatyer 1977) and some experimental studies (e.g. McCormick 1968, Hils and Vankat 1982) have revealed situations in which these classical facilitation processes do not occur, even though there may be a pattern of replacement of species over time. However, the description of a pattern of species change (i.e. 'physiognomic change' or 'a parade of predominants'; Egler 1981) as a succession has a subtle influence on many readers who anticipate some of the processes that comprise the facilitation model even if these are not implied by an author. Because succession theory developed strongly in North America with such clear examples as the studies of glacial retreats (Crocker and Major 1955), sand dune stabilization (Olson 1958) and revegetation of abandoned farmland, it is particularly important to guard against the application of expected *processes* to a familiar *pattern* that emerges after different sorts of disturbance (i.e. fire) and in different regions.

These warnings about the application of succession terminology to classify post-fire community changes are perhaps most appropriate in the fire-prone ecosystems of the world, where many species have evolved to tolerate, in some way, the passage of the fire, enabling them to retain some occupancy of the site. Hence, removal of species and invasion of propagules are not the predominant processes occurring to produce community change. There have been critical examinations about succession in fire-prone ecosystems for a long time, focussing on several components of succession theories, such as: (i) whether there is likely to be vegetation change towards some predictable climax community (see the delightful account of a discussion at a 1950 meeting of the British Ecological Society – Clapham 1950); (ii) whether post-fire vegetational changes are successional, resulting in new definitions such as 'autosuccession' for Californian chaparral (Hanes 1971), 'cyclical succession', (Watt 1947, Barclay-Estrup and Gimingham 1969), or simply rejections of the facilitation (relay floristics) model (e.g. Purdie and Slatyer 1976, Whelan and Main 1979); and (iii) whether the nature of the community that develops is set solely or mostly by processes occurring in a brief window of time just after fire (see Holland 1986).

Processes causing community change

Clearly it is more profitable to focus directly on the *processes* that might produce observed patterns of community changes after fires (Table 7.4), than to infer them from an apparent fit of the pattern of post-fire community change to some model or other. It will be apparent from the list of examples in Table 7.4 that many of the changes in communities are the result of interactions between the characteristics of the fire itself, processes that occur after fire such as climatic conditions and herbivory, and survival and dispersal attributes of the organisms in relation to the fire and to the post-fire processes. Thus, the approach of defining 'vital attributes' of species in a community (e.g. Noble and Slatyer 1977) has potential to predict community changes, as long as the complexity of interactions can be incorporated. For example, a low-intensity fire may be patchy in coverage, thus reducing the importance of differential seed dispersabilities as a factor, but increasing the importance of selective herbivory in the burned patches.

Earlier sections of this chapter have touched on the operation of many of these processes in explaining the community's response to fire. For example, long-term changes in forest tree composition in northern hemisphere conifer forests may be explained by the relative susceptibilities of tree species to a particular fire and the relative dispersabilities of

Table 7.4. *Processes that can produce changes in communities after fires. Any of these processes that act with differing magnitudes on different species may make the post-fire community differ from the pre-fire community and may produce continuing change*

Process	Examples of modes of operation
Survival of fire by adult plants	Western hemlock eliminated from Douglas fir forests, conifers persist
	Obligate-seeder shrubs disfavoured by frequent fires – favoured by fires of intermediate frequency
Survival of fire by propagules on site	Closed-cone species (e.g. black spruce in Alaska) recolonize rapidly while other species must reinvade (e.g. white spruce)
	Californian post-fire ephemerals have long-term dormancy between fires
Dispersal of seeds into site	Wind-dispersed species such as *Epilobium* occupy sites first after fire, followed by more poorly dispersed species
Post-fire climate	Post-fire climate is more significant for recovery of populations of obligate-seeder species than sprouters
Herbivory	Selective post-fire herbivory may eliminate mass-germinating populations of *Acacia* after fire
	Reinvasion of herbivores some time after fire may affect plants that became established soon after the fire
	Pre-dispersal seed predation can reduce seed banks of species with on-plant or underground seeds, especially if a long time elapses since last fire
Competition	Later invaders such as hardwoods in Florida sandhills may outcompete pines in the absence of further burning
Other interactions	Invasion of certain plant species may require prior establishment of appropriate symbiotic fungi

their propagules. One process that has been much neglected, however, though it has great potential to direct changes in community structure in relation to fire and after other sorts of disturbance, is herbivory (Edwards and Gillman 1987, Whelan 1989).

Post-fire herbivory and community composition
Exclosure studies, which test the effects of herbivores on the development of a plant community after fire, reveal that while herbivores may have little effect on altering plant community structure in the absence of fire, the combination of fire and herbivory can cause a marked change (see Leigh and Holgate 1979; Chapter 5).

In an experimental study in North American prairie, Collins (1987) showed that grazing and burning both had effects on the species richness and species composition of the grassland community, but their greatest impact was in combination (Fig. 7.24). Cattle grazing alone did not have a great influence on species diversity; burning alone causes a reduction in species diversity; both disturbances together caused a marked increase in diversity.

The results of Collins (1987) lend support to the theory that disturbances such as fire and grazing can maintain high species diversity (see below). However, increased species diversity as a result of grazing, burning, or both, is not universal by any means (Whelan 1977, 1989). Burning can make otherwise long-lived plants susceptible to herbivory, thus reducing diversity by the elimination of palatable species. A clear example of this is the history of vegetation change on Rottnest Island – off the coast of south-western Australia. At the turn of the twentieth century, the vegetation of this island was dominated by dense thickets of tall, woody shrubs and trees, including *Acacia rostellifera*, *Callitris preissii* and *Melaleuca lanceolata* (Pen and Green 1983). The area of dominance of these tall woody species has declined from about 1000 ha in 1919 to less than 150 ha. The current dominant is a low, prickly shrub, *Acanthocarpus preissii*. The demise of the tall shrub species, *A. rostellifera*, is attributable to intense herbivory by a small marsupial macropod, the quokka (*Setonix brachyurus*), which occurs after fires that were patchy and/or small in area. Exclosures erected by A.R. Main (pers. commun.) after a wildfire illustrate the potential for *Acacia* recruitment in the absence of herbivory by the marsupials (Fig. 7.25)

Using caging experiments in southern Californian chaparral, Mills (1986) showed that survivorship of ceanothus seedlings (*Ceanothus greggii*) was reduced much more than *Adenostoma fasciculatum* (chamise)

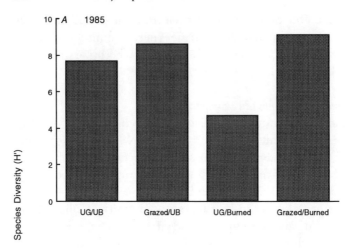

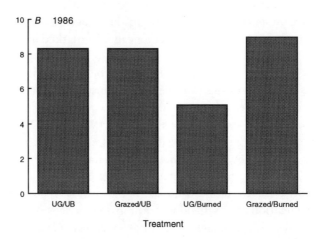

Figure 7.24 The interaction between grazing and fire can influence species diversity more than either factor alone. In North American prairie, species diversity was lowest, over two years – 1985 (*A*) and 1986 (*B*), for sites that were burned but not grazed and highest for sites that were both burned and grazed. UG = ungrazed; UB = unburned (data from Collins 1987).

by herbivores, over 3 years after a fire. Mills predicted that this pattern of preferential grazing on ceanothus was likely to determine the relative abundances of these two chaparral dominants in the post-fire community.

It is notable that insect herbivores have been largely ignored in relation to fire, especially those that may have an impact below ground. Studies by Brown and co-workers (e.g. Brown 1985, Brown and Gange

Figure 7.25 Current vegetation of Rottnest Island, in Western Australia, dominated by *Acanthocarpos preissii* (background), is maintained by an interaction between fire and grazing by small marsupials. Patches of vegetation fenced after fire support dense *Acacia rostellifera* thickets (foreground).

1989, Brown *et al.* 1988, Hendrix *et al.* 1988), illustrate the potential for experimental removals of insects to examine their role in plant community development after disturbance. Using insecticide treatments in abandoned agricultural land at Silwood Park, England, Hendrix *et al.* (1988) demonstrated that insects were reducing the species richness of annual plants in the first year after abandonment. By the second year, insects were retarding the growth of perennial grasses, compared with insecticide-treated plots, thereby increasing species richness overall but slowing the rate at which the expected successional process to perennial grassland developed.

It must be concluded that there is enormous potential for herbivory to control the patterns of plant species composition developing after fire, yet few studies have really addressed this hypothesis.

Complex interactions
Herbivory is relatively simple to study compared with other sorts of processes that may interact with fire to determine community composition. The challenge of untangling the effects of numerous potentially interacting factors is still to be taken up, especially in relation to soil processes, such as symbioses with mycorrhizal fungi. Too little is known about how these sorts of processes might influence the post-fire community to make a review meaningful, though there are some suggestions that further research here will be most profitable. Bowman (1993) found some unusual patterns in survivorship of monsoon forest and savanna tree species planted across a fire-boundary, and he speculated (D. Bowman pers. commun.) that pathogenic soil microorganisms or symbiotic mycorrhizal fungi may be removed by fire.

Studies by Christensen (1980), Malajczuk *et al.* (1987), Taylor (1991, 1992), Claridge (1992) and Claridge *et al.* (1992) suggest some complex interactions between marsupials, mycorrhizal fungi and plants in relation to fire. Fire apparently increases the density of sporocarps of the dominant fungal genus (*Mesophellia*; Taylor 1991). Small marsupials such as woylies (*Bettongia* species) in south-western Australia and potoroos (*Potorous* species) in south-eastern Australia depend on these sporocarps as a major food item, thus dispersing fungal spores. It is not clear how crucial the animals are in dispersing and 'processing' spores, as fires may not eliminate the mycorrhizae from a site, although Claridge *et al.* (1992) showed that colonization of eucalypt seedlings followed experimental application of *Potorous* faeces but not direct application of the fungus. How would the recovery of the plant community be influenced if a fire removed all marsupials from a site? Does animal activity increase the probability that roots of a new seedling will intercept fungal spores? Does reinvasion of particular plant species to a site after fire depend upon the prior dispersal to the site of appropriate mycorrhizal fungi?

Resilience, stability and species diversity

Resilience and stability

The enormous variation in vegetation responses to fire in different regions, with some communities hardly changing after fire and others changing enormously, has inevitably led to the development of more abstract, community-level concepts, such as 'resilience'. Is a Mediterranean shrubland more resilient, when exposed to fires, than a north-eastern temperate forest (see B. J. Fox and M. D. Fox 1986)? Can estimates of resilience, if they may be made with any accuracy, help predict the effects of human interference in natural ecosystems, as suggested by Westman (1986)?

Westman (1986) defined a number of terms which, if they could be measured, might permit comparisons of ecosystems in different regions, and which might aid in the prediction of the possible effects of altered fire regimes. These include *inertia* – the degree of resistance to change when some stress (e.g. fire) is applied to the ecosystem, and *resilience* – the rate and manner of return to an original condition after the stress has been applied. The following components of resilience may be identified (Westman 1986):

Elasticity	rate of recovery of some ecosystem property after disturbance;
Amplitude	the threshold levels of ecosystem alteration beyond which it will not return to its original state;
Hysteresis	the degree of similarity between the ecosystem change under chronic disturbance and the manner of recovery afterwards;
Malleability	the degree to which the post–disturbance steady-state differs from the original condition.

It is easy to identify problems in using these concepts to explain or predict post-fire vegetation change, or the response of a vegetation to an altered fire regime. What ecosystem property is most relevant, and can it be readily measured (Grubb and Hopkins 1986)? Is it possible to quantify a steady state or pre-disturbance state of an ecosystem that is frequently being influenced by fire?

At present, the main value of these concepts appears to be to concentrate our attention in the patterns of change in ecosystems after an alteration of conditions. In relation to the concept of elasticity, the following questions are relevant, and we would focus on the effects of a single fire. Does species richness recover more rapidly after fire than after other disturbances in Mediterranean-climate ecosystems (e.g. B. J. Fox and M. D. Fox 1986)? Does richness return more rapidly after fire in Mediterranean-climate systems, where seed dispersal from off-site is not an important process in vegetation recovery, than in closed forests, where adults die and seeds must be re-introduced from the boundaries of the site? Does biomass or cover recover more rapidly on coastal than inland sites? B. J. Fox and M. D. Fox (1986) attempted to address the resilience question by an examination of the shapes of species-richness curves – in different ecosystems – for a variety of different disturbances (old-fields, post-fire, post-mining, grazing). Their review illustrated that existing studies are inadequate to permit any conclusion about resiliance, damping and other similar processes.

In relation to the concept of amplitude, we could focus on the effects of a chronically applied disturbance (i.e. a changed fire regime). What change in fire regime would be required to shift a community to a new equilibrium state? Perhaps this is most clearly illustrated by considering the invasion of fire-prone woodlands and forests by hardwoods and closed-forest species in the long absence of fire, as discussed above. With a high fire frequency in the appropriate time of year, hardwood species may be held at bay, as consecutive fires kill recruits that make it into the surrounding forest. If, however, fire was absent for some time, the invading hardwoods could alter the fuel conditions and slowly replace the pines, thereby making future fires less and less likely (Platt and Schwartz 1990).

Could a single fire displace a community beyond the threshold of return to its current equilibrium? This, of course, would depend on the fire, and there are some examples of marked vegetation change producing an apparently stable, new community, rather than a successional change. For example, Arseneault and Payette (1992) described a site in northern Canada that was a lichen–spruce krummholz before a fire in 1750. Since 1750, this site has apparently been occupied by a lichen–tundra vegetation community. It is likely that changes to new, stable communities after a single fire would occur only after particularly severe fires, or fires in regions that have rarely or never experienced fires before.

Landscape-level species diversity
The question of scale has been implicit in much of the above discussion. Should species richness be measured in 1 m^2, 100 m^2 or hectare plots? Do small patches of burned vegetation respond to fire in the same way as a large, burned landscape? This chapter started with a statement that the community was an appropriate level of organization to be used by land managers, and that species richness was one emergent property that could be relatively easily monitored as an indicator of change. Cowling (1987) and Keith (1991) have argued that effective management requires a knowledge of the effects of fire on species richness at a larger scale.

The theoretical framework for understanding the control of species diversity is competitive exclusion and coexistence. What processes might minimize competitive exclusion and thereby maximize coexistence? There are two sets of hypotheses: equilibrium and non-equilibrium. Equilibrium hypotheses argue that a system has reached equilibrium and competitive interactions have maximum opportunity to operate. The remaining species richness reflects the subdivision of the available resources among the remaining species. The non-equilibrium hypotheses argue that some process such as disturbance prevents the system from reaching equilibrium, i.e. competitive interactions between species are relaxed and coexistence can occur even if species have the same patterns of resource use.

Fire can be a disturbance that fends off competitive exclusion. This could occur at a number of spatial scales. For example, recurring fires at low frequency in a central Florida pineland may operate to maintain populations of pines and other xeric species by preventing hardwoods from achieving dominance. Thus, a particular site may always have a mixture of these types of species. Alternatively, competitive processes may occur vigorously over time after fire, as expected in the facilitation model of plant succession, such that species richness at any one site may be quite low, but the degree of heterogeneity of patches of vegetation of differing post-fire ages in the landscape will result in a high species richness at the landscape level.

Various studies have implicated fire in the production and maintenance of high species diversity, especially in heathlands growing on nutrient-poor soils of south-western Australia and South Africa (e.g.

Hopper 1979, Cowling 1987, Bond *et al.* 1992). The mechanisms include the reduction of competitive dominance by established individuals and provision of recruitment opportunities for a range of species, which may occupy different 'safe sites' by chance (lottery hypotheses), by differing regeneration niches (Grubb 1977), or by successive fires favouring recruitment of different suites of species (e.g. Bergl and Lamont 1988, Pierce and Cowling 1991, Moreno and Oechel 1991, Smith *et al.* 1992). Spatial heterogeneity provided by fires has been invoked to explain the maintenance of high species diversity of lizards in Australian deserts (Pianka 1992).

Outstanding questions

The above discussion of changes in communities in relation to fire reveals gaps in our knowledge, many of which are attributable to the difficulty of carrying out large-scale, replicated, manipulative experiments with fire regime as the manipulated variable. The most important challenge for the future is therefore to attempt large-scale manipulations of fire regime, with relevant measures of community change and a facility for detecting the effects of interacting factors.

The following can be identified as some of the important research needs.

1. Experimental studies of changes in community parameters such as species composition with replication of fires.
2. Experiments manipulating components of fire regime (e.g. fire frequency, fire season) over long time spans, and measuring changes in species diversity and composition.
3. Detailed studies on dispersal of propagules of a range of plant species into burned sites, in a range of regions, to elucidate the degree to which species reappearance after fire comes from outside the perimeter of the burned area.
4. The inclusion of more than a single trophic level in experimental studies, especially experimental studies on the role of herbivores, including insects, on post-fire plant communities.
5. Focus on the role of below-ground interactions, such as symbiotic mycorrhizae.
6. A thorough review of plant succession theory as it relates to fire ecology, comparing processes responsible for vegetation changes in regions with different fire histories.

7. Investigations of the importance of 'specific conditions' on the nature of community changes after fire. For example, can post-fire climatic condition have an over-riding influence on the post-fire relative abundance of plant species (say obligate seeders *versus* sprouters), making differential pre-fire relative abundances or differential build-up of seed banks less important.

8. Studies of the relative importance of factors such as food or cover in the recovery of animal communities after fire.

8 · *Fire and management*

Fire in vegetation: a bad master, a good servant and a national
problem   *(Phillips (1930))*

No harsher criticism was ever lodged against the old Division of Forestry
than Gifford Pinchot's strictures on its failure to practice forestry in the
woods. He chided his predecesor's administration (1886–1898) for
confining its attention to research and the propagation of information
while eschewing responsibility for applying its findings to the formidable
task of forest management.   *(Schiff (1962))*

Introduction

Fire has long been used as a tool in land management. There is much
written on the past and current use of fire by indigenous peoples. Good
examples come from various accounts of Australian Aboriginal use of
fire in their custodianship of the 'bush' (e.g. Pyne 1991a,b). Deliberate
fires were used to clear dense vegetation to make travel easier, to flush
game from dense vegetation, to attract game back into regrowth, to
regenerate plant speces that were suitable as foods and to provide
firebreaks around campsites (Jones 1969).

To present-day land managers also, fire represents a relatively cheap
land management tool. Controlled fire can be used to protect against
subsequent wildfire, to maintain particular species of plants or animals,
to eliminate undesirable species, to enhance species diversity, and for
numerous other reasons. The first of these uses, namely using fire to
reduce the likelihood of occurrence and the intensity of a subsequent
wildfire, has received the most attention, driven predominantly by the
needs of forestry to protect the timber resource.

What do land managers need to know about the effects of fire on the
ecology of the system for which they are responsible? What information
presented in the preceding chapters of this book is most relevant to
decision-making by land managers? Is there already a concise body of

ecological knowledge available upon which to base predictions of the consequenses of applying particular management policies? Reading the previous chapters should make it quite clear that there is not nearly sufficient ecological knowledge available – and this must surely be the greatest challenge for land managers. The theme with which this book concludes is therefore '... every management activity is an experiment and should be treated as such' (Preece 1990; see also Christensen *et al.* 1989). The following list (see Preece 1990) succinctly summarizes a number of problems with the interface between ecological theory and land management: problems to which many land managers will relate:

● Many land managers lack appropriate scientific training; even if they possess it, they lack the time to interpret the data that are available and to translate them into practice;
● Most researchers have a limited understanding of the complications of applying theory to practice and of identifying the data that are of most relevance to management;
● Ecological theories are poorly developed, they are too general to be of use in particular situations, and are still being debated by ecologists;
● Expertise in applying particular fire regimes to an area has resided in very few people;
● There are few reliable data available on the effectiveness of prescribed burning in achieving particular management aims.

The problems raised by these issues have been succinctly put by McAnninch and Strayer (1989):

... managers often need some immediate answers, while scientists often are unwilling or unable to provide complete answers without a lengthy study. A central problem here is the difficulty in dealing with scientific uncertainty. Often scientists are unwilling to provide scientific advice on a problem unless there is a very high degree of scientific certainty about the solutions being offered. On the other hand, a manager may use scientific advice and knowledge as if it were the final solution, and be unwilling to alter management practices and regulations in light of changing conditions and revised scientific ideas.

The main purpose of this chapter is to identify areas of ecology, discussed in the previous part of the book, which are likely to be of particular relevance in land management, and to point to some strategies which could be adopted by managers in an attempt to overcome some of the problems summarized in Preece's list, above.

Defining the objectives of management

Clearly the main objectives of land management, in a particular area, in relation to fire must be articulated carefully, because they will determine the sort of ecological knowledge that is required and the sort of concerns which should be held. Controlled use of fire can be an integral part of a wide range of land managment activities, from maintaining populations of a single endangered species to maximizing runoff in a water catchment area (Table 8.1).

It could be argued that, once a decision had been made to manage an area for a particular objective (e.g. to maximize water runoff, to enhance the forage potential of an area used for rough grazing), the ecological considerations of the effects of fires prescribed to achieve the objective become irrelevant. There are two counter-arguments, however. First, whether the fire regime prescribed will actually achieve the management objective may depend on the ecological effects of the fires, especially on the flora. For example, frequent fires combined with intense grazing may lead to losses in the palatable plant species that supported grazing. Second, there is seldom only a single value associated with a particular area. For example, protection of vegetation cover in a water catchment area may be a primary management objective, but species conservation may be a secondary one (Whelan and Muston 1991).

This is the principle of multiple use, which has frequently been put forward, especially by forestry industries. Although suspiciously regarded as a rationalization to obtain access to as much forested land as possible, despite conservation values in the land, the idea of multiple use is becoming increasingly prevalent. Recent national and international strategies for the preservation of biological diversity (e.g. the World Biodiversity Summit in Rio de Janeiro – June 1992; DEST 1992) have explicitly recognized the need for *ex situ* conservation, that is, the contribution to conservation of biodiversity which must be made by lands outside a national park and nature reserve system. It will become increasingly important to ensure that land management activities, such as prescribed burning, satisfy a conservation objective as well as more utilitarian objectives.

Hazard-reduction burning

The use of deliberate fire to prevent high-intensity wildfires has become probably the most extensive use of fire in land management. Protection

Table 8.1. *Various management objectives which may involve the deliberate use of fire, and which may therefore require some knowledge of the ecological effects of an applied fire regime*

Management objective	Use of fire in achieving objective
Forestry	Rotational hazard-reduction burning to prevent widespread crown fires
	Removal of species in competition with desired timber species
	Control of soil-dwelling pathogens (e.g. *Phytophthora*) and weeds
	Stimulation of regeneration of desired tree species by high-intensity
Flower harvesting	Maximization of production of inflorescences on woody perennials, particularly the Proteaceae in Africa
Water resources	Maximization of runoff without erosion
Primary production	Stimulation of 'green pick' for stock in rangeland
	Removal of pathogens and parasites of livestock
Urban	Low fuel loads around installations/ subdivisions, etc.
National Parks, etc.	Protection of installations/neighbours
	Maintenance of particular species/communities that require a specific fire regime (includes decision to allow wildfires burn out)
	Control of soil-dwelling pathogens (e.g. *Phytophthora*) and weeds
	Creation of wildflower displays

Source: Modified from Edwards (1984) and van Wilgen *et al.* (1992).

of the forestry resource has been the primary impetus for the development of hazard-reduction techniques in at least two continents – North America and Australia (Whelan and Muston 1991). The events leading up to this were the same in both continents: (i) a legacy of the northern European attitude that high-intensity fire must be bad produced stringent fire-suppression policies, identified in North America with Smokey the Bear; (ii) consequent build-up of fuel loads; (iii) inevitable wildfires, when they occurred, were uncontrollable, extensive conflagrations; (iv) development of techniques of reducing the hazard of a wildfire, predominantly by frequent 'burning-off' in large blocks of land on rotation (Good 1981).

Successful fire suppression in the early part of the twentieth century led inevitably to steadily increasing fuel loads over an increasing proportion of the landscape. Nevertheless, in the south-eastern USA there was resistance to changing the policies of complete fire suppression, even in the face of experimental evidence which showed that productivity in southern pine forests was actually increased by periodic controlled burning (Schiff 1962). Schiff's book, *Fire and Water*, is an almost unbelievable tale of suppression and delay of research results which contradicted the policy of fire suppression. It took large-scale wildfires to change opinion, both in the USA in the 1940s and in Australia in the 1950s (Schiff 1962, Shea *et al.* 1981), with the result that forest departments developed strategies for extensive hazard-reduction burning to prevent future catastrophic fires. Opinion changed in forest departments because these high-intensity fires caused economic losses to the industry by damaging trees, and frequent hazard-reduction burning demonstrably caused less damage to established trees. Hence, the primary management objective, namely protection of a timber resource from damage by high-intensity fire, was satisfied.

The development of hazard-reduction strategies was based on a knowledge of fire behaviour in particular forest types (see Chapter 2). In Australia, this research was pioneered by McArthur (1962), and in North America, by Byram (1959), van Wagner (1990) and Rothermel (1972). As discussed in Chapter 2, fuel load is a major contributor to fire intensity and rate of spread. Hence, hazard-reduction burning is designed to be frequent enough to maintain fuel loads at a level below which they would not support an uncontrollable fire. This may mean that a given block of vegetation would be burned as frequently as every 5 years, if productivity is high – less frequently for sites with low productivity and hence a slow rate of accumulation of fuel.

Forest departments appear to have been so good at devising hazard-reduction burning techniques and publicizing the need for them, that the prescription has been adopted in many land management decisions, without a close examination of whether this use of fire might be compromising the management objectives (see Good 1981, Whelan and Muston 1991). Even in forest industry lands, identified for wood production, there may be other management objectives related to multiple use which are compromised by frequent, low-intensity fire in the cool season of the year. Williams *et al.* (1994) referred to this situation of blind adoption of a management practice in one area following its successful use elsewhere as 'portability of prescription'. Clearly, a management prescription is not portable without testing at the new site.

Using a set of case studies described in Whelan and Muston (1991), the following section examines a range of potential problems associated with the application of a burning program designed for reduction of fuels, using regular, frequent fires usually set in a cool season of the year, when control is easiest.

Problems with fire regimes imposed for fuel reduction

Local extinction of obligate seeders

The life-history of the obligate-seeder plant species was described in Chapters 3 and 5. Although fires kill established plants, these species respond to fire by germination from a stored seed bank (stored either in the soil or in woody fruits in the canopy). Populations of these species will be at risk if a second fire occurs before the post-fire flush of seedlings has had an opportunity to develop a seed bank of its own. Suggestions that local extinctions do indeed occur with high fire frequencies come from studies by Nieuwenhuis (1987) on relative abundances of re-sprouters and obligate seeders in south-eastern Australian woodlands, described in Chapter 7, and by Zedler *et al.* (1983) on the effects of two closely consecutive fires in Californian chaparral, described in Chapter 2.

Predicting the effects of increased fire frequency in an area will require knowledge of the life-histories of obligate-seeder species, including particularly the intensity of fire likely to kill established plants, and the age of first reproduction of seedlings. These characteristics are likely to vary among sites, as shown for age of first reproduction in eastern Australian shrub species (Benson 1985: see Chapter 5, Fig. 5.8). More-over, several factors may provide some resilience for populations of obligate seeders, in the face of frequent fires: firstly, spatial heterogeneity in fuels produced by topographic discontinuities will provide refugia unlikely to be burned in consecutive fires; and secondly, high fire frequency is likely to produce fires that are of low intensity and patchy, because fuel loads do not have sufficient time to build up and become continuous (Whelan and Muston 1991).

Introduction of a weed-dominated community

The connection between the fire regime in an area and encouragement of weed encroachment has been recognized in many parts of the world, especially with the spread of Australian *Acacia* and *Hakea* in southern Africa (van Wilgen *et al.* 1990) and Australian *Melaleuca* in south Florida (Wade *et al.* 1980, Myers and Ewel 1990). Cochrane (1963, 1969) argued that increased fire frequency at the Mount Lofty Ranges in southern Australia encouraged the establishment of introduced weeds which produced a greater volume of more flammable fuel more quickly than the native species which they replaced. With frequent opportunity for ignition of fires, the new fuel characteristics would sustain the more-frequent fire regime and cause even greater changes to the original community.

Faunal habitat

Protection of faunal habitat is one common reason for the setting up of reserves. Conservation of fauna is also often one of the important secondary values in areas otherwise used for forestry, water catchment, etc. It is possible that a fire regime which is effective in reducing the danger from wildfires also reduces the value of the habitat for fauna.

An example of this is the protection of populations of two threatened bird species in south-eastern Australia: the ground parrot (*Pezoporus wallicus*) and the eastern bristle bird (*Dasyornis brachypterus*). These species both inhabit areas of sedgeland, and are distributed in relatively few small, disjunct populations on the eastern Australian mainland (Blakers *et al.*1984, Meredith *et al.* 1984). These sedgelands provide a continuous fuel which is flammable even in moist conditions, and firefighters recall fighting fires while standing ankle-deep in water! Moreover, the fuel accumulates rapidly as most of the sedgeland species sprout from lignotubers and other rootstocks. Hence, to be really effective in protecting against wildfires, a rotational burning program would have to ensure that each block of vegetation is burned about every three years. It is difficult to determine the responses of ground parrot populations to fire, but it is apparent that population densities do not peak for many years after fire (Meredith *et al.* 1984, R. Jordan pers. commun.), perhaps well over ten – depending upon the site (Baker and Whelan 1994). Therefore, achieving the hazard-reduction objective is likely to cause a continued decline in the parrot population. Eastern bristle birds favour multi-layered eucalypt forests and woodlands, in the forest–sedgeland mosaic. Frequent hazard-reduction burning is likely to convert these multi-layered forested patches to bilayered communities, with a eucalyt canopy and heath–sedge ground layer: a profile which is not favoured by bristle birds (J. Baker, unpublished data).

Season of burning

Hazard-reduction burning is typically carried out at a time of year when it is easiest to control the fires: in Mediterranean-climate regions, this is therefore winter and spring. The frequency distribution of occurrence of wildfires in these regions peaks in late summer and autumn (see Chapter 2). Concerns are frequently voiced, especially by conservation organizations, about the potentially deleterious ecological impact of out-of-season burning, yet there are few data available. Van Wilgen *et al.* (1990) expressed concerns based on plant population dynamics: the timing of fruit maturation in fynbos plant communities providing many viable seeds after a fire in autumn, but few after a fire in spring (see also Chapter 5). Moreover, Bond (1984) showed that seeds released from fruits in response to a spring fire had to remain on the soil surface for months, exposed to predation until conditions appropriate for germination occurred. Seeds released after an autumn fire were able to germinate virtually immediately, thus escaping predation (see Fig. 3.21).

Animal populations are also likely to be affected by the season of burning, with timing of aestivation, nesting and mating all making individuals and populations more susceptible to fires in some seasons than others (see Chapter 6).

Area and patchiness of burned sites

The typical hazard-reduction burning programme produces fires which do not burn all of the vegetation in an area. In fact, the objective of hazard-reduction burning may be to achieve as little as 40% burned. Burning in the cool season or during weather conditions cool and damp enough to ensure adequate control is virtually certain to produce a mosaic of burned and unburned vegetation. Aerial ignition in particular produces this sort of pattern, as there are multiple ignitions which do not have an opportunity to develop into a continuous fire front (see Plate 1 of Shea *et al.* 1981). In addition to the pattern of burned and unburned vegetation within a block of vegetation subjected to a hazard-reduction fire, the blocks themselves are often small in area.

A consequence of these patterns is that animals are easily able to escape the fire fronts, and unburned vegetation provides refuges from which herbivores can move back into the burned patches to forage (see the study of small marsupials by Christensen 1980, described in Chapter 5). In some areas, management of vegetation includes deliberate alternation of burned and unburned patches to maximize populations of herbivores, for example the management of heather moorlands for grouse (Lovat 1911, Hobbs and Gimingham 1987). These responses of herbivores to patchy fires suggest that these fires may well result in overgrazed vegetation and hence loss of palatable plant species. Such vegetation changes due to post-fire herbivory were described in Chapters 5 and 7.

In contrast to the problem of increased herbivory in small, patchy fires, is the situation where mammalian herbivory may be severely reduced because of wildfire in island-like National Parks that are surrounded by agriculture and/or urban areas (see Le Maitre 1992). A fire such as the 1994 wildfire that burned about 90% of Sydney's Royal National Park may well have severely reduced or even eliminated populations of some mammal species. The vegetation is likely to change with reduced herbivory, just as it would with increased herbivory, as described above. Reinvasion of mammals in Royal National Park will have to occur through two rather narrow corridors of vegetation that link this park with nearby large areas of forest (see Robinson 1977) – the rest of the park bordering ocean or the Sydney metropolitan area.

Fire intensity and seed germination

Seeds of many species, particularly legumes, remain dormant in the soil, to be stimulated to germinate by the heat of a fire (see Chapter 3). As heat penetration downwards through soil is poor, the peak temperature and duration of heating by the fire will determine the amount of germination at a particular depth of burial. Cool fires will produce germination only from the upper layer of soil, while an intense fire may kill seeds near the surface and stimulate germination of those buried deeper. Several studies report that low-intensity fires, such as typical cool-season, hazard-reduction fires, may fail to produce dense germination of legumes. Localized hot spots may produce patches of good germination, but these may be particularly susceptible to herbivory, as reported by Shea *et al.* (1979).

Summary

In summary, these case studies indicate that the adoption of cool-season, frequent, hazard-reduction burning may carry with it undesirable ecological side-effects. These side-effects, if they are known, can be weighed up against the primary management objective for a site – be it maintenance of water quality and quantity, protection of a timber resource from wildfire, or protection of property from wildfire. It is clear that, if preservation of biodiversity (populations or communities in a biota) is the primary management objective, great care should be taken before a prescription of frequent, cool-season burning is adopted. It is also obvious that there is unlikely to be sufficient ecological information available to be certain of the ecological effects of *any* prescribed fire regime. Hence, management will have to be experimental. In addition, putting all eggs in one basket, by imposing the same, regular fire regime across a landscape, will be an unwise strategy. It is unlikely that all objectives for land in multiple use will be able to be achieved under one fire regime.

Prescribing fire regimes for ecological reasons

Every major wildfire that threatens human lives and property is followed by accusations about inadequate hazard reduction. Many fire control agencies, politicians and members of the general public believe that protection from high-intensity wildfires could be assured if only all areas of natural vegetation had been 'burned off' frequently enough. Following the serious fires in Sydney in January 1994, which burned homes in the metropolitan area and burned most of Australia's oldest National Park (Royal National Park), there were strident calls for a halt to the declaration of any new National Parks or Wilderness Areas and for large-scale hazard-reduction burning of any large tract of natural vegetation. Hazard-reduction burning at sufficient frequency to ensure wildfire control during extreme weather conditions may have to be very frequent indeed in some vegetation types. A thorough evaluation of these issues is beyond the scope of this book. However, the previous section indicates clearly that broad-scale, fuel-reduction burning, if it is frequent, would compromise the principal management objectives of many areas of natural vegetation.

This argument is not to say that deliberate burning should not be prescribed in areas designated for conservation. On the contrary – there are many ecological reasons for prescribing fire, but what fire regime should be prescribed?

Mimicking the historical regime

Where preservation of biodiversity is the principal management objective for an area, what is the appropriate fire regime to apply? It has been argued that an historical fire regime should be mimicked, based on the reasoning that an 'unnatural' fire regime is certain to produce ecological problems because elements of the biota have evolved under a different set of fire conditions. This may be a reasonable argument, but it is very difficult to decide just what the historical fire regime was. Reconstructions of long-term fire histories are not good, even for post-settlement times. Although fire scars in tree rings may provide information on the historical frequency of high-intensity fires, a great deal of effort is required to provide the intensity of sampling required to gain information on the spatial extent of fires, and season of fires is virtually impossible to infer. Interpretation of pollen and charcoal in sediments suffers from a similar set of problems. Anthropological evidence of patterns of burning by aboriginal peoples pertains to relatively recent times, and how relevant this burning will have been to the evolution of characteristics of the biota is a matter of argument.

In large areas of natural vegetation, one management strategy which may be used to maintain historic fire regimes is 'let wildfires burn'. Few reserves are now large and isolated enough to allow this management strategy to be adopted without a great deal of public criticism, as illustrated by the great debates following the 1988 wildfires in the Greater Yellowstone Area (Christensen *et al.* 1989, Schullery 1989). Christensen *et al.* (1989) came to the following conclusion about the 'let wildfires burn' strategy:

When, following the Leopold Report (1963) and the Wilderness Act (1964), the dilemma of fire protection in wilderness and parks became an object of intense scrutiny, many observers believed that wilderness fire management involved merely the restoration of a natural process into a natural environment.... In its purest expression, a wilderness fire program sought to withdraw suppression practices and permit naturally ignited fires to burn freely. Since then, the philosophical issues have blurred and the operational problems have multiplied. It is not enough to withdraw aggressive suppression from wild areas. Rather, fire management must blend various forms of suppression with various forms of prescribed fire.

In the light of the constraints on our ability to infer an historical fire regime in an area, uncovering the limits of tolerance of the biota to an altered fire regime is an important approach. Kilgore (1973) argued that, in management, we should be attempting to produce the range of fire *effects* found historically, rather than attempting to discover and mimic fire season, intensity, frequency and size of burns. Robbins and Myers (1992) used this principle in constructing the following list of issues that should be considered when deciding on an appropriate fire regime to prescribe for a natural area.

1. What is the natural response of the target ecosystem(s) to the fire regimes to which it might be exposed?
2. What are the fire and land-use histories of the site? What environmental and cultural factors produced the present site conditions?
3. What are the life-histories of the species of concern (endangered, keystone, dominant, indicator)? How do they respond to different fire regimes?
4. What are the community dynamics and how do community structures vary under different fire regimes?
5. What components or processes of the ecosystem are missing or irreversibly altered (e.g. lack of intercommunity connections, species extirpations or extinctions, exotic species introductions)?
6. How important is the seasonal component of fire to each of the above? Can other facets of fire regime be manipulated to achieve the same result?
7. Can a reasonable facsimile of the natural fire regime be created? What compromises need to be made for safety and public relations, or because of conflicting uses?

It should be reiterated that there will be two main reasons for burning: (i) to reduce the likelihood of uncontrollable wildfires; (ii) to provide conditions necessary for the biota (see Christensen *et al.* 1989). The important question in relation to the first reason for burning is: 'What departure from the 'natural' regime will the biota tolerate?'. In relation to the second reason for burning, the important questions are usually: 'What fire regime is necessary?' and 'Can it be achieved?'.

Van Wilgen *et al.* (1990) illustrated clearly that we can expect difficulties in imposing appropriate fire regimes, because the window of time available between the months of high fire danger and the months in which it may be ecologically unacceptable to burn may be quite small (Fig. 8.1). Hence, there may be only a few weeks per year available for large areas to be burned as part of a management programme.

Prescribing high intensity fire

The Tall Timbers Fire Ecology Conference in 1989 (Hermann *et al.* 1991) addressed the problems of prescribing high-intensity fires. Increasing dissection of natural landscapes by roads, towns, agriculture, etc. has brought with it increased public pressure to reduce the likelihood of uncontrollable wildfires, and increased legal constraints on deliberate burning. The need for high-intensity fire to maintain some types of plant

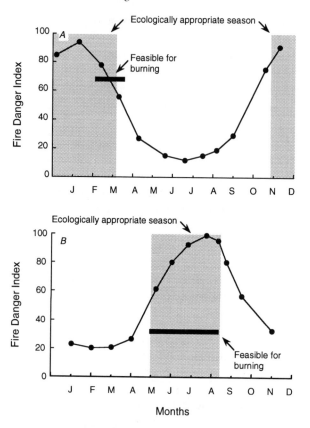

Figure 8.1 The annual cycle of fire danger in (*A*) fynbos, and (*B*) grasslands in South Africa (from van Wilgen *et al.* 1990), illustrating that the time of year in which prescription burning is acceptable for ecological reasons (stippled) generally coincides with the season of highest fire danger.

community was discussed in Chapter 7. A good example is the maintenance of sand pine scrub in south Florida (Myers 1985, 1990, 1991; Fig. 7.9). This community is the habitat for many endangered or threatened species, all of which appear to depend on a regime of infrequent but high-intensity fire (Myers 1991). However, the small size of scrub remnants, dissection by roads from other communities and the proximity of development such as condominiums, citrus groves and residential subdivisions will demand innovative approaches to fire management, reserve design and forms of manipulating vegetation by means other than fire.

Similar problems, and therefore a similar need for innovations in management strategies, occur in urban nature reserves in many fire-prone areas of the world. It is possible to achieve appropriate management using high-intensity fire in these situations. Wade (1991) described the successful use of high-intensity fire in a challenging management problem – maintaining a *Spartina* marsh in a small urban reserve with a board-walk and visitors centre constructed within the marsh! Whelan and Muston (1991) reported on an attempt, in south-western Australia, to solve the problem of congregation of herbivores following small-area management fires. Temporary electric fences were used to reduce, for the first year, the numbers of kangaroos using the relatively small areas which were burned.

Research and monitoring

It is evident that the recipes for hazard-reduction burning, developed for the protection of particular tree species in commercial forestry, may be readily transferred to other land areas and there is considerable and constant pressure to do this, especially after each episode of wildfire adjacent to centres of human habitation. The material presented in this book indicates that potential ecological effects of imposing a particular fire regime, especially one that differs from the historic regime, are both complex and difficult to predict. This complexity and uncertaintly makes it difficult to argue against the widespread adoption of hazard-reduction prescriptions. It is easier to focus on the technological and legal problems of smoke control and the need to protect neighbouring property than to decide on the 'right' fire regime for an area based on ecological objectives. Unless the ecological research is addressed, however, we risk modifying many plant communities and compromising management objectives in many areas.

As discussed in Chapters 4 and 7, long-term experimental studies are badly needed, but are still rare, partially because they are so difficult to set up, maintain and monitor. Management cannot leave the experimental work solely to researchers and many authors recently have emphasized the need to include experimental work and monitoring as an integral part of management (e.g. Gill 1983, Christensen and Maisey 1987, Christensen *et al.* 1989, Preece 1990, Whelan and Muston 1991).

Land managers are annually planning prescribed burning for large areas of land and are therefore in an ideal situation to establish large-scale experiments. However, it is apparent that competing demands on the time and resources available to land managers is increasingly constraining the possibility for research and monitoring. This trend must be reversed, and there is an urgent need for explicit collaborations between land management authorities and researchers in research institutions.

There are several major challenges for ecological research in relation to land management, based on the establishment and monitoring of long-term, replicated, experimental studies to test various ecological effects of fire, in particular:

(i) The effects of different frequencies and seasons of burning on plant and animal communities, focussing both on loss of particular species and alterations to plant associations;

(ii) The effects of the area of burn and patchiness on processes such as reinvasion of wind-dispersed plant species and especially the impact of herbivory;

(iii) The testing of predictions based on population dynamics models (see Chapter 5), which necessarily contain simplifying assumptions (see Whelan and Muston 1991).

Finally, if there is a successful collaboration between researchers and managers, there is an urgent need to see the results communicated in the mainstream ecological literature. Much of the information obtained in experimental management and in ecological monitoring after management will be of great value in the develoment of ecological theory in relation to fire, and the results of management must therefore not be buried in the 'grey literature'.

References

Abbott, I. (1984a) Changes in the abundance and activity of certain soil and litter fauna in the jarrah forest of Western Australia after a moderate intensity wildfire. *Australian Journal of Soil Research* **22**: 463–469.

Abbott, I. (1984b) Emergence, early survival, and growth of seedlings in six tree species in Mediterranean forest of Western Australia. *Forest Ecology and Management* **9**: 51–66.

Abbott, I. (1985) Reproductive ecology of *Banksia grandis* (Proteaceae). *New Phytologist* **99**: 129–148.

Abbott, I. & Loneragan, O. (1983) Influence of fire on growth rate, mortality and butt damage in Mediterranean forest of Western Australia. *Forest Ecology and Management* **6**: 139–153.

Abbott, I., van Heurck, P. & Wong, L. (1985) Responses to long-term fire exclusion: physical, chemical and faunal features of litter and soil in a Western Australian forest. *Australian Forestry* **47**: 237–242.

Abrahamson, W. G. & Hartnett, D. C. (1990) Pine flatwoods and dry prairies. In *Ecosystems of Florida*, ed. Myers, R. L. & Ewel, J. J., pp. 103–149. Florida: Academic Press.

Adams D. E. & Anderson, R. C. (1978) The response of a central Oklahoma grassland to burning. *Southwestern Naturalist* **23**: 623–632.

Adamson, D., Selkirk, P. M. & Mitchell, P. (1983) The role of fire and lyre-birds in the landscape of the Sydney Basin. In *Aspects of Australian Sandstone Landscapes*, ed. Young, R. W. & Nanson, G. L., pp. 81–93. Australia: University of Wollongong.

Adamson, R. S. (1935) The plant communities of Table Mountain. III. A six years study of regeneration after burning. *Journal of Ecology* **23**: 43–55.

Ahlgren, C. E. (1966) Small mammals and reforestation following prescribed burning. *Journal of Forestry* **64**: 614–618.

Ahlgren, I. F. (1974) The effect of fire on soil organisms. In *Fire and Ecosystems*, ed. Kozlowski, T. T. & Ahlgren, C. E., pp. 47–72. New York: Academic Press.

Ahlgren, I. F. & Ahlgren, C. E. (1960) Ecological effects of forest fires. *Botanical Reviews* **26**: 483–533.

Alexander, M. E. (1982) Calculating and interpreting forest fire intensities. *Canadian Journal of Botany* **60**: 349–357.

Alexandrov, V. Y. (1964) Cytophysiological and cytoecological investigations of heat resistance of plant cells toward the action of high and low temperature. *Quarterly Review of Biology* **39**: 35–77.

Allaby, M. (1985) *The Oxford Dictionary of Natural History*. Oxford: Oxford University Press.

Allee, W. C., Emerson, A. E., Park, O., Park, T. & Schmidt, K. P. (1949) *Priciples of Animal Ecology*. London: Saunders.

Andersen, A. N. (1988) Immediate and longer term effects of fire on seed predation by ants in sclerophyllous vegetation in south-eastern Australia. *Australian Journal of Ecology* **13**: 285–293.

Andrew, M. H. & Mott, J. J. (1983) Annuals with transient seed banks: the population biology of indigenous *Sorghum* species of tropical north-west Australia. *Australian Journal of Ecology* **8**: 265–276.

Andrews, E. F. (1917) Agency of fire in propagation of Longleaf Pines. *Botanical Gazette* **64**: 497–508.

Antonovics, J. & Levin, D. A. (1980) The ecological and genetic consequences of density dependent regulation in plants. *Annual Review of Ecology and Systematics* **11**: 411–452.

Arata, A. A. (1959) Effects of burning on vegetation and rodent populations in a Longleaf Pine Turkey Oak association in North Central Florida. *Journal of the Florida Academy of Science* **22**: 94–104.

Arseneault, D. & Payette, S. (1992) A postfire shift from lichen–spruce to lichen–tundra vegetation at tree line. *Ecology* **73**: 1067–1081.

Ashton, D. H. (1970) The effects of fire on vegetation. *Proceedings of the 2nd Fire Ecology Symposium*, Melbourne: Monash University.

Ashton, D. H. (1976) The development of even-aged stands of *Eucalyptus regnans* F. Muell. in Central Victoria. *Australian Journal of Botany* **24**: 397–414.

Ashton, D. H. (1979) Seed harvesting by ants in forests of *Eucalyptus regnans* F. Muell. in Central Victoria. *Australian Journal of Ecology* **4**: 265–277.

Ashton, D. H. (1981) The ecology of the boundary between *Eucalyptus regnans* F. Muell. and *E. obliqua* L'Herit. in Victoria. *Proceedings of the Ecological Society of Australia* **11**: 74–94.

Ashton, D. H. (1986) Viability of seeds of *Eucalyptus obliqua* and *Leptospermum juniperinum* from capsules subjected to a crown fire. *Australian Forestry* **49**: 28–35.

Athias-Binche, F. (1987) Regeneration patterns of Mediterranean ecosystems after fire: the case of some soil arthropods. 3. Uropodid mites. *Vie et Milieu* **37**: 39–52.

Auld, T. D. (1984) *Population Dynamics of Acacia suaveolens* (Sm.) Willd. Unpublished PhD Thesis, Australia: University of Sydney.

Auld, T. D. (1986a) Dormancy and viability in *Acacia suaveolens* (Sm.) Willd. *Australian Journal of Botany* **34**: 463–472.

Auld, T. D. (1986b) Population dynamics of the shrub *Acacia suaveolens* (Sm.) Willd.: dispersal and the dynamics of the soil seed bank. *Australian Journal of Ecology* **11**: 235–254.

Auld, T. D. (1987) Population dynamics of the shrub *Acacia suaveolens* (Sm.) Willd.: survivorship throughout the life cycle, a synthesis. *Australian Journal of Ecology* **12**: 139–151.

Auld, T. D. & Myerscough, P. J. (1986) Population dynamics of the shrub *Acacia suaveolens* (Sm.) Willd.: seed production and predispersal seed predation. *Australian Journal of Ecology* **11**: 219–234.

Austin, M. P. (1985) Continuum concept, ordination methods, and niche theory. *Annual Review of Ecology and Systematics* **16**: 39–61.

Austin, M. P. (1990) Community theory and competition in vegetation. In *Perspectives on Plant Competition*, ed. Grace, J. B. & Tilman, D., pp. 215–238. New York: Academic Press.

Bailey, A. W. & Anderson, M. L. (1980) Fire, temperatures in grass, shrub, and aspen forest communities of Central Alberta. *Journal of Rangeland Management* 3: 37–40.

Baird, A. M. (1977) Regeneration after fire in King's Park, Perth, Western Australia. *Journal of the Royal Society of Western Australia* 60: 1–22.

Baker, J. & Whelan, R. J. (1994) Ground Parrots and fire at Barren Grounds, New South Wales. A long-term study and an assessment of management implications. *Emu* 94: 300–304.

Ballardie, R. T. & Whelan, R. J. (1986) Masting, seed dispersal and seed predation in the cycad *Macrozamia communis*. *Oecologia* 70: 100–105.

Bamber, R. K. & Mullette, K. J. (1978) Studies of the lignotubers of *Eucalyptus gummifera* (Gaertn. & Hochr.). II Anatomy. *Australian Journal of Botany* 26: 15–22.

Barclay-Estrup, P. & Gimingham, C. H. (1969) The description and interpretation of cyclical processes in the heath community. I. Vegetational change in relation to the *Calluna* cycle. *Journal of Ecology* 57: 737–758.

Barden, L. S. (1979) Serotiny and seed viability of *Pinus pungens* in the southern Appalachians. *Castanea* 44: 44–47.

Beadle, N. C. W. (1940) Soil temperatures during forest fires and their effect on the survival of vegetation. *Journal of Ecology* 28: 180–192.

Beaton, M. J. (1982) Fire and water: Aspects of Australian Aboriginal management of cycads. *Archaeology in Oceania* 17: 59–67.

Beaufait, W. R. (1960) Some effects of high temperatures on the cones and seeds of Jack Pine. *Forest Science* 6: 194–199.

Beaufait, W. R. (1961) Crown temperatures during prescribed burning in Jack Pine. *Papers of the Michigan Academy of Science, Arts & Letters* LVI: 251–257.

Beckwith, R. C. and Werner, R. A. (1979) Effects of fire on arthropod distribution. In *Ecological Effects of the Wickersham Dome Fire Near Fairbanks, Alaska*, ed. Viereck, L. A. & Dyrness, C. T., pp 53–55. USDA General Technical Report PNW-90. Oregon: Portland.

Begon, M. & Mortimer, M. (1981) *Population Ecology: A Unified Study of Animals and Plants*. (1st edn.) Massachusetts: Sinauer.

Begon, M. & Mortimer, M. (1986) *Population Ecology: A Unified Study of Animals and Plants*. (2nd edn.). Massachusetts: Sinauer.

Begon, M., Harper, J. L. & Townsend, M. (1986) *Ecology: Individuals, Populations, Communities*. Oxford: Blackwell Scientific.

Begon, M., Harper, J. L. & Townsend, M. (1990) *Ecology: Individuals, Populations, Communities*, 2nd edn. Oxford: Blackwell Scientific.

Bell, D. T. (1985) Aspects of response to fire in the northern sandplain heathlands. In *Symposium on Fire Ecology and Management in Western Australian Ecosystems*, ed. J. R. Ford, pp. 33–40. Western Australia: School of Biology, Curtin University.

Bell, D. T. & Koch, J. M. (1980) Post-fire succession in the northern jarrah forest of Western Australia. *Australian Journal of Ecology* 5: 9–14.

Bell, D. T., Hopkins, A. J. M. & Pate, J. S. (1984) Fire in the Kwongan. In *Plant Life on the Sandplain*, ed. Pate, J. S. & Beard, J. S., pp. 178–204. Nedlands, Western Australia: University of Western Australia Press.

Bell, D. T., Vlahos, S. & Watson, L. E. (1987) Stimulation of seed germination of understorey species of the northern Jarrah forest of Western Australia. *Australian Journal of Botany* **35**: 593–599.

Bellgard, S. E., Whelan, R. J. & Muston, R. M. (1994) The impact of wildfire on vesicular–arbuscular mycorrhizal fungi and their potential to influence the re-establishment of post-fire plant communities. *Mycorrhiza* **4**: 139–146.

Bendell, J. F. (1974) Effect of fire on birds and mammals. In *Fire and Ecosystems*, ed. Kozlowski, T. T. & Ahlgren, C. E., pp. 73–138. New York: Academic Press.

Benson, D. H. (1985) Maturation periods for fire-sensitive shrub species in Hawkesbury Sandstone vegetation. *Cunninghamia* **1**: 339–349.

Bentley, J. R. & Fenner, R. L. (1958) Soil temperatures during burning related to postfire seedbeds on woodland range. *Journal of Forestry* **56**: 737–740.

Berenstain, L. (1986) Responses of long-tailed Macaques to drought and fire in eastern Borneo: a preliminary report. *Biotropica* **18**: 257–262.

Bergeron, Y. & Brisson, J. (1990) Fire regime in red pine stands at the northern limit of the species' range. *Ecology* **71**: 1352–1364.

Bergl, S. M. & Lamont, B. B. (1988) The water relations, rooting patterns and phenologies of two sclerophyllous shrubs. In *Proceedings of the Fifth International Conference on Mediterranean Ecosystems*, pp. 569–573. International Union of Biological Sciences, Paris.

Bewley, J. D. & Black, M. (1982) The release from dormancy. In *Physiology and Biochemistry of Seeds in Relation to Germination*, vol. 2, pp. 184–198. Berlin: Springer-Verlag.

Bigalke, R. C. & Pepler, D. (1979) Unpublished records. Department of Nature Conservation, University of Stellenbosch.

Bigalke R. C. & Willan K. (1984) Effects of fire regime on faunal composition and dynamics. In *Ecological Effects of Fire in South African Ecosystems*, ed. Booysen, P. de V. & Tainton, N. M., pp. 255 -271. New York: Springer-Verlag.

Billings, W. D. (1938) The structure and development of oldfield shortleaf pinestands and certain associated properties of the soil. *Ecological Monographs* **8**: 437–499.

Biswell, H. H. (1974) Effects of fire on chaparral. In *Fire and Ecosystems*, ed. Kozlowski, T. T. & Ahlgren, C. E., pp. 321–364. New York: Academic Press.

Blaisdell, J. P. & Mueggler, W. F. (1956) Sprouting of bitterbush (*Purshia tridentata*) following burning or top removal. *Ecology* **37**: 365–370.

Blakers, M., Davies, S. J. J. F. & Reilly, P. N. (eds.) (1984) *The Atlas of Australian Birds*. Melbourne: Melbourne University Press.

Bock, C. E. & Bock, J. H. (1978) Response of birds, small mammals and vegetation to burning sacaton grasslands in southeastern Arizona. *Journal of Rangeland Management* **31**: 296–300.

Bock, J. H., Bock, C. E. & McKnight, J. R. (1976) A study of the effects of grassland fires at the research ranch in southeastern Arizona. *Arizona Academy of Science* **11**: 49–57.

Bond, W. J. (1984) Fire survival of Cape Proteaceae: influence of fire season and seed predators. *Vegetatio* **56**: 65–74.

Bond, W. J. (1985) Canopy-stored seed reserves (serotiny) in Cape Proteaceae. *South African Journal of Botany* **51**: 181–186.

Bond, W. J., Vlok, J. & Viviers, M. (1984) Variation in seedling recruitment of Cape Proteaceae after fire. *Journal of Ecology* **72**: 209–221.

Bond, W. J., Cowling, R. M. & Richards, M. B. (1992) Competition and coexistence. In *The Ecology of Fynbos: Nutrients, Fire and Diversity*, ed. Cowling, R., pp. 206–225. Oxford: Oxford University Press.

Bond, W. J., Vlok, J. & Viviers, M. (1984) Variation in seedling recruitment of Cape Proteaceae after fire. *Journal of Ecology* **72**: 209–221.

Booysen, P. de V. & Tainton, N. M. (eds.) (1984) *Ecological Effects of Fire in South African Ecosystems*. New York: Springer-Verlag.

Bowman, D. M. J. S. (1993) Establishment of two dry monsoon forest tree species on a fire-protected monsoon forest–savanna boundary, Cobourg Peninsula, northern Australia. *Australian Journal of Ecology* **18**: 235–237.

Bowman, D. M. J. S. & Fensham, R. J. (1991) Response of a monsoon forest–savanna boundary to fire protection, Weipa, northern Australia. *Australian Journal of Ecology* **16**: 111–118.

Bowman, D. M. J. S. & Jackson, W. D. (1981) Vegetation succession in southwest Tasmania. *Search* **12**: 358–362.

Bowman, D. M. J. S., Wilson, B. A. & Hooper, R. J. (1988) Response of *Eucalyptus* forest and woodland to four fire regimes at Munmarlary, Northern Territory, Australia. *Journal of Ecology* **76**: 215–232.

Boyer, D. E & Dell, J. D. (1980) *Fire Effects on Pacific Northwest Soils*. USDA Forest Service, Pacific Northwest Region. Oregon: Portland.

Bradstock, R. A. & Bedward, M. (1992) Simulation of the effect of season of fire on post-fire seedling emergence of two *Banksia* species based on long-term rainfall records. *Australian Journal of Botany* **40**: 75–88.

Bradstock, R. A. & Myerscough, P. J. (1981) Fire effects on seed release and the emergence and establishment of seedlings in *Banksia ericifolia* L.f. *Australian Journal of Botany* **29**: 521–531.

Bradstock, R. A. & O'Connell, M. A. (1988) Demography of woody plants in relation to fire: *Banksia ericifolia* L.f. and *Petrophile pulchella* (Schrad) R. Br. *Australian Journal of Ecology* **13**: 505–518.

Braithwaite, R. W. (1990) A new savannah fire experiment. *Bulletin of the Ecological Society of Australia* **20**: 47–48.

Brakefield, P. M. (1987) Industrial melanism: do we have the answers? *Trends in Ecology and Evolution* **2**: 117–122.

Bramble, W. C. & Goddard, M. K. (1942) Effect of animal coaction and seedbed condition on the regeneration of Pitch Pine in the Barrens of central Pennsylvania. *Ecology* **23**: 330–335.

Brewer, J. S. & Platt, W. J. (1994) Effects of fire season and herbivory on reproductive success in a clonal forb, *Pityopsis graminifolia* (Michx.) Nutt. *Journal of Ecology* **82**: 665–675.

Broughton, S. K. and Dickman, C. R. (1991) The effect of supplementary food on home range of the southern brown bandicoot, *Isoodon obesulus* (Marsupialia: Peramelidae). *Australian Journal of Ecology* **16**: 71–78.

Brown, V. K. (1985) Insect herbivores and plant succession. *Oikos* **44**: 17–22.

Brown, V. K. & Gange, A. C. (1989) Herbivory by soil-dwelling insects depresses plant species richness. *Functional Ecology* **3**: 667–671.

Brown, V. K., Jepsen, M. & Gibson, C. W. D. (1988) Insect herbivory: effects on early old field succession, demonstrated by chemical exclusion methods. *Oikos* **52**: 293–302.

Bruce, D. (1954) Mortality of Longleaf Pine seedlings after a winter fire. *Journal of Forestry* **52**: 442–443.

Buckley, R. (1983) The role of fire in maintaining a small-scale vegetation gradient: a test of Mutch's hypothesis. *Oikos* **41**: 291–292.

Buckley, R. (1984) The role of fire: response to Snyder. *Oikos* **43**: 405–406.

Buffington, J. D. (1967) Soil arthropod populations of the New Jersey pine barrens as affected by fire. *Annals of the Entomological Society of America* **60**: 530–535.

Burbidge, A. H. & Whelan, R. J. (1982) Seed dispersal in the cycad *Macrozamia riedlei*. *Australian Journal of Ecology* **7**: 63–67.

Burgan, R. E. & Rothermel, R. C. (1984) BEHAVE: Fire behavior prediction and fuel modeling system – fuel subsystem. *USDA Forest Service General Technical Report* INT-167.

Burrough, P. A., Brown, L. & Morris, E. C. (1977) Variations in vegetation and soil patterns across the Hawkesbury Sandstone Plateau from Barren Grounds to Fitzroy Falls, New South Wales. *Australian Journal of Ecology* **2**: 137–159.

Burrows, F. M. (1986) The aerial motion of seeds, fruits, spores and pollen. In *Seed Dispersal*, ed. Murray, D. R., pp. 1–47. Sydney: Academic Press.

Burrows, N. D. (1985) Reducing the abundance of *Banksia grandis* in the jarrah forest by the use of controlled fire. *Australian Forestry* **48**: 63–70.

Burtt, E. (1951) The ability of adult grasshoppers to change colour on burnt ground. *Proceedings of the Royal Entomological Society of London* **26**: 45–49.

Byram, G. M. (1959) Combustion of forest fuels. In *Forest Fire: Control and Use*, ed. Davis, K. P., pp. 61–89. New York: McGraw-Hill.

Cable, D. R. (1973) Fire effects in southwestern semidesert grass–shrub communities. *Proceedings of the Tall Timbers Fire Ecology Conference* **12**: 109–127.

Campbell, A. J. & Tanton, M. T. (1981) Effects of fire on the invertebrate fauna of soil and litter of a eucalypt forest. In *Fire and the Australian Biota*, ed. Gill, A. M., Groves, R. H. & Noble, I. R., pp. 213–241. Canberra: Australian Academy of Science.

Cancelado, R. & Yonke, T. R. (1970) Effect of prairie burning on insect populations. *Journal of the Kansas Entomological Society* **43**: 274–281.

Carpenter, F. L. & Recher, H. F. (1979) Pollination, reproduction and fire. *American Naturalist* **113**: 871–879.

Carreira, J. A., Sanchez-Vazquez, F. & Niell, F. X. (1992) Short-term, and small-scale patterns of post-fire regeneration in a semi-arid dolomitic basin of southern Spain. *Acta Oecologia* **13**: 241–253.

Carrol, E. J. & Ashton, D. H. (1965) Seed storage in soils of several Victorian plant communities. *Victorian Naturalist* **82**: 102–112.

Cary, A. (1932) Some relations of fire to longleaf pine. *Journal of Forestry* **30**: 594–601.

Catchpole, T. & de Mestre, N. (1986) Physical models for a spreading line fire. *Australian Forestry* **49**: 102–111.

Catling, P. C. & Newsome, A. E. (1981) Responses of the Australian vertebrate fauna to fire: an evolutionary approach. In *Fire and the Australian Biota*, ed. Gill, A.

M., Groves, R. H. & Noble, I. R., pp. 273–310. Canberra: Australian Academy of Science.

Caughley, G., Brown, B. & Noble, J. (1985) Movement of kangaroos after a fire in mallee woodland. *Australian Wildlife Research* **12**: 349–353.

Cavers, P. B. (1983) Seed demography. *Canadian Journal of Botany* **61**: 3578–3590.

Chaiken, L. E. (1952) Annual summer fires kill hardwood root stocks. *USDA Forest Service South East Forest Experimental Station* Research Note 19.

Chandler, C., Cheney, P., Thomas, P., Trabaud, L. & Williams, D. (1983) *Fire in Forestry*. Volumes 1 & 2. New York: Wiley.

Chapin, F. S. & van Cleve, K. (1981) Plant nutrient absorption and retention under differing fire regimes. In *Fire Regimes and Ecosysytem Properties*, ed. Mooney, H. A., Christensen, N. L., Lotan, J. E. & Reiners, W. A., pp. 303–324. *USDA General Technical Report* WO-26.

Chapman, H. H. (1932) Is the longleaf pine type a climax? *Ecology* **13**: 32–334.

Chattaway, M. M. (1958) The regenerative powers of certain eucalypts. *Victorian Naturalist* **75**: 45–46.

Cheney, N. P. (1981) Fire behaviour. In *Fire and the Australian Biota*, ed. Gill, A. M., Groves, R. H. & Noble, I. R., pp. 151–175. Canberra: Australian Academy of Science.

Cheplick, G. P. & Quinn, J. A. (1988) Subterranean seed production and population responses to fire in *Amphicarpum purshii* (Gramineae). *Journal of Ecology* **76**: 263–273.

Chew, R., Butterworth, B. & Grechman R. (1958) The effects of fire on the small mammal population of chaparral. *Journal of Mammalogy* **40**: 253.

Christensen, N. L. (1981) Fire regimes in southeastern ecosystems. In *Fire Regimes and Ecosystem Properties*, ed. Mooney, H. A., Christensen, N. L., Lotan, J. E. & Reiners, W. A., pp. 112–136. *USDA General Technical Report* WO-26.

Christensen, N. L. & Muller, C. H. (1975) Effects of fire on factors controlling plant growth in *Adenostoma* chaparral. *Ecological Monographs* **45**: 29–55.

Christensen, N. L., Agee, J. K., Brussard, P. F., Hughes, J., Knight, D. H., Minshall, G. W., Peek, J. M., Pyne, S. J., Swanson, F. J., Thomas, J. W., Wells, S., Williams, S. E. & Wright, H. A. (1989) Interpreting the Yellowstone fires of 1988: ecosystem responses and management implications. *Bioscience* **39**: 678–685.

Christensen, P. E. S. (1978) The concept of fauna priority areas. *Proceedings of the 3rd Fire Ecology Symposium*. Forestery Commission of Victoria & Monash University.

Christensen, P. E. S. (1980) The biology of *Bettongia penicillata* Gray, 1837, and *Macropus eugenii* (Demarest, 1817) in relation to fire. *Forests Department of Western Australia* Bulletin 91.

Christensen, P. E. S. & Annels, A. (1985) Fire in southern tall forests. In *Symposium on Fire Ecology and Management in Western Australian Ecosystems*, ed. Ford, J. R., pp. 67–82. Nedlands, Western Australia: Botany Department, University of Western Australia.

Christensen, P. E. S. & Kimber, P. C. (1975) Effect of prescribed burning on the flora and fauna of south-west Australian forests. *Proceedings of the Ecological Society of Australia* **9**: 85–106.

Christensen, P. E. S. & Maisey, K. (1987) The use of fire as a management tool in fauna conservation reserves. In *Nature Conservation: The Role of Native Vegetation*, ed. Saunders, D. A., Arnold, G. W., Burbidge, A. A. & Hopkins, A. J. M., pp. 323–329. Sydney: Surrey Beatty & Sons.

Christensen, P. E. S., Wardell-Johnson, G. & Kimber, P. (1985) Birds and fire in southwestern forests. In *Birds of Eucalypt Forests and Woodlands*, ed. Keast, A., Recher, H. F., Ford, H. & Saunders, D., pp. 291–299. Sydney: Surrey Beatty & Sons.

Christian, D. P. (1977) Effects of fire on small-mammal populations in a desert grassland. *Journal of Mammalogy* **58**: 423–427.

Clapham, A. R. (1950) London meeting: Thursday, 30 March 1950. *Journal of Ecology* **38**: 412.

Claridge, A. W. (1992) Is the relationship among mycophagous marsupials, mycorrhizal fungi and plants dependent on fire? *Australian Journal of Ecology* **17**: 223–225.

Claridge, A. W., Tanton, M. T., Seebeck, J. H., Cork, S. J. & Cunningham, R. B. (1992) Establishment of ectomycorrhizae on the roots of two species of *Eucalyptus* from fungal spores contained in the faeces of the long-nosed potoroo (*Potorous tridactylus*). *Australian Journal of Ecology* **17**: 207–217.

Clark, R. L & Wasson, R. J. (1986) Reservoir sediments. In *Limnology in Australia*, ed. De Decker, P. and Williams, W. D., pp. 497–507. Australia: CSIRO & Dordrecht: Junk.

Clark, S. S. (1988) Effects of hazard-reduction burning on populations of understorey plant species on Hawkesbury sandstone. *Australian Journal of Ecology* **13**: 473–484.

Clayton-Greene, K. A. and Beard, J. S. (1985) The fire factor in vine thicket and woodland vegetation of the Admiralty Gulf region, north-west Kimberly, Western Australia. *Proceedings of the Ecological Society of Australia* **13**: 225–230.

Cochrane, G. R. (1963) Vegetation studies in forest-fire areas of the Mount Lofty Ranges, South Australia. *Ecology* **44**: 41–52.

Cochrane, G. R. (1969) Fire ecology in southeastern Australian sclerophyll forests. *Proceedings of the Tall Timbers Fire Ecology Conference* **8**:15–39.

Cody, M. L. & Mooney, H. A. (1978) Convergence versus nonconvergence in Mediterranean-climate ecosystems. *Annual Review of Ecology and Systematics* **9**: 265–321.

Colgan, N. (1913) Further notes on the flora of burnt ground. *The Irish Naturalist* **22**: 85–93.

Collins, S. L. (1987) Interaction of disturbances in tallgrass prairie: a field experiment. *Ecology* **68**: 1243–1250.

Connell, J. H. (1980) Diversity and the coevolution of competitors, or the ghost of competition past. *Oikos* **35**: 131–138.

Connell, J. H. & Slatyer, R. O. (1977) Mechanisms of succession in natural communties and their role in community stability and organisation. *American Naturalist* **111**: 1119–1144.

Cook, S. F. (1959) The effects of fire on a population of small rodents. *Ecology* **40**: 102–108.

Cope, M. J. & Chaloner, W. G. (1985) Wildlife: an interaction of biological and physical processes. In *Geological Factors and the Evolution of Plants*, ed. Tiffney, B. H., pp. 257–277. Connecticut: Yale University Press.

Copland, B. J. & Whelan, R. J. (1989) Seasonal variation in flowering intensity and pollination limitation of fruit set in four co-occurring *Banksia* species. *Journal of Ecology* **77**: 509–523.

Cowling, R. M. (1987) Fire and its role in coexistence and speciation in Gondwanan shrublands. *South African Journal of Science* **83**: 106–112.

Cowling, R.(ed.) (1992) *The Ecology of Fynbos: Nutrients, Fire and Diversity.* Oxford: Oxford University Press.

Cowling, R. M. & Lamont, B. B. (1984) Population dynamics and recruitment of four co-occurring *Banksia* spp. after spring and autumn burns. In *Medecos IV*, ed. Dell, B., pp. 31–32. Nedlands, Western Australia: Botany Department, University of Western Australia.

Cowling, R. M. & Lamont, B. B. (1985a) Seed release in *Banksia*: the role of wet–dry cycles. *Australian Journal of Ecology* **10**: 169–171.

Cowling, R. M. & Lamont, B. B. (1985b) Variation in serotiny of three *Banksia* species along a climatic gradient. *Australian Journal of Ecology* **10**: 345–350.

Cowling, R. M. & Lamont, B. B. (1987) Post-fire recruitment of four co-occurring *Banksia* species. *Journal of Applied Ecology* **24**: 645–658.

Cowling, R. M., Lamont, B. B. & Pierce, S. M. (1987) Seed bank dynamics of four co-occurring Banskia species. *Journal of Ecology* **75**: 289–302.

Cremer, K. W. (1973) Ability of *Eucalyptus regnans* and associated evergreen hardwoods to recover from cutting or complete defoliation in different seasons. *Australian Forest Research* **6**: 9–22.

Crocker, R. L. & Major, J. (1955) Soil development in relation to vegetation and surface age at Glacier Bay, Alaska. *Journal of Ecology* **43**: 427–448.

Crowner, A. W. & Barrett, G. W. (1979) Effects of fire on the small mammal component of an experimental grassland community. *Journal of Mammalogy* **60**: 803–813.

Curtis, J. T. & Partch, M. L. (1950) Some factors affecting flower production in *Andropogon gerardi*. *Ecology* **31**: 488–489.

Cushwa, C. T., Martin, R. E. & Miller, R. L. (1968) The effects of fire on seed germination. *Journal of Range Management* **21**: 250–254.

Cypert, E. (1961) The effects of fires in the Okefenokee Swamp in 1954 and 1955. *American Midland Naturalist* **66**: 485–503.

Daubenmire, R. (1968) Ecology of fire in grassland. *Advances in Ecological Research* **5**: 209–266.

DeBano, L. F. (1971) The effect of hydrological substances on water movement in soil during infiltration. *Proceedings of the Soil Science Society of America* **35**: 340–343.

DeBano, L. F., Dunn, P. H. & Conrad, C. E. (1977) Fire's effects on physical and chemical properties of chaparral soils. In *Environmental Consequences of Fire and Fuel Management in Mediterranean Ecosystems*, pp. 65–74. USDA Forest Service General Technical Report WO-3.

DeBano, L. F., Osborn, J. F., Krammes, J. S. & Letey, J. (1967) Soil wettability and wetting agents . . . our current knowledge of the problem. *USDA Forest Service Research Paper* PSW-43.

DeBano, L. F., Savage, S. M. & Hamilton, D. M. (1976) The transfer of heat and hydrophobic substances during burning. *Journal of the Soil Science Society of America* **40**: 779–782.

Deevey, E. S. (1947) Life tables for natural populations of animals. *Quarterly Review of Biology* **22**: 283–314.

Department of the Environment, Sport and Territories (DEST) (1992) *Biological Diversity: Its Future Conservation in Australia.* Canberra: DEST.

DeWitt, J. B. & Derby, J. V. (1955) Changes in nutritive value of browse plants following forest fires. *Journal of Wildlife Management* **19**: 65–70.

Dhillion, S. S., Anderson, R. C. & Liberta, A. E. (1988) Effect of fire on the mycorrhizal ecology of little bluestem (*Schizachyrium scorparium*). *Canadian Journal of Botany* **66**: 706–713.

Diamond, J. & Case, T. J. (1986) *Community Ecology.* New York: Harper & Row.

Dickman, C. R. (1989) Demographic responses of *Antechinus stuartii* (Marsupialia) to supplementary food. *Australian Journal of Ecology* **14**: 387–398.

Dieterich, J. H. & Swetnam, T. W. (1984) Dendrochronology of a fire-scarred Ponderosa Pine. *Forest Science* **30**: 238–247.

Dolva, G. (1993) The effect of fire on the ecology and life-history of the wood cricket *Nambungia balyarta* (Nemobiinae: Gryllidae). Unpublished Masters Thesis, University of Western Australia.

Drury, W. H. & Nisbett, I. C. T. (1973) Succession. *Journal of the Arnold Arboretum* **54**: 331–368.

Ebersole, J. J. (1989) Role of the seed bank in providing colonizers on a tundra disturbance in Alaska. *Canadian Journal of Botany* **67**: 466–471.

Edgell, M. C. R. & Brown, E. H. (1975) The bushfire environment of southeastern Australia. *Journal of Environmental Management* **3**: 329–349.

Edwards, C. A. & Fletcher, K. E. (1971) A comparison of extraction methods for terrestrial arthropods. In *Methods of Study in Quantitative Soil Ecology: Population, Production and Energy Flow*, ed. Phillopson, J., pp. 150–185. Oxford: Blackwell Scientific.

Edwards, P. J. (1984) The use of fire as a management tool. In *Ecological Effects of Fire in South African Ecosystems*, ed. Booysen, P.de V. & Tainton, N. M., pp. 349–362. Berlin: Springer-Verlag.

Edwards, P. J. & Gillman, M. P. (1987) Herbivores and plant succession. In *Colonization, Succession and Stability*, ed. Gray, A. J., Crawley, M. J. & Edwards, P. J., pp. 295–310. Oxford: Blackwell Scientific.

Egler, F. E. (1954) Vegetation science concepts. I. Initial floristic composition, a factor in old-field vegetation development. *Vegetatio* **14**: 412–417.

Egler, F. E. (1981) Letters to the editor. *Bulletin of the Ecological Society of America* **62**: 230–232.

Enright, N. J. & Lamont, B. B. (1989) Seed banks, fire season, safe sites and seedling recruitment in five co-occurring *Banksia* species. *Journal of Ecology* **77**: 1111–1122.

Erickson, R. (1965) *Orchids of the West.* Perth, Western Australia: Paterson Brokensha.

Euler, D. L. & Thompson, D. Q. (1978) Ruffed Grouse and songbird foraging response on small spring burns. *New York Fish and Game Journal* **25**: 156–164.

Evans, E. W., Rogers R. A. & Opfermann, D. J. (1983) Sampling grasshoppers

(Orthoptera: Acrididae) on burned and unburned tallgrass prairie: night trapping vs. sweeping. *Environmental Entomology* **12**: 1449–1454.

Ewing, A. L. & Engle, D. M. (1988) Effects of late summer fire on tallgrass prairie microclimate and community composition. *American Midland Naturalist* 120: 212–223.

Fahnestock, G. R. & Hare, R. C. (1964) Heating of tree trunks in surface fires. *Journal of Forestry* **62**: 779–805.

Floyd, A. G. (1966) Effect of fire upon weed seeds in the wet sclerophyll forests of northern New South Wales. *Australian Journal of Botany* **14**: 243–256.

Floyd, A. G. (1976) Effect of burning on regeneration from seeds in wet sclerophyll forest. *Australian Forestry* **39**: 210–220.

Force, D. C. (1981) Postfire insect succession in southern California chaparral. *American Naturalist* **117**: 575–582.

Foster, T. (1976) *Bushfire: History, Prevention, Control.* Sydney: Reed.

Fox, B. J. (1981) The influence of disturbance (fire, mining) on ant and small mammal species diversity in Australian heathland. In *Proceedings of the Symposium on Dynamics and Management of Mediterranean-Type Ecosystems*, pp. 213–219. USDA Forest Service General Technical Report PSW-58.

Fox, B. J. (1982) Fire and mammalian secondary succession in an Australian coastal heath. *Ecology* **63**: 1332–1341

Fox, B. J. (1983) Mammals species diversity in Australian heathlands: the importance of pyric succession and habitat diversity. In *Mediterranean-type Ecosystems: The Role of Nutrients*, ed. Kruger, F. J., Mitchell, D. T. & Jarvis, J. U. M., pp. 473–489. Berlin: Springer-Verlag.

Fox, B. J. & Fox, M. D. (1978) Recolonization of coastal heath by *Pseudomys novaehollandiae* (Rodentia: Muridae) following sand mining. *Australian Journal of Ecology* **3**: 447–465.

Fox, B. J. & Fox, M. D. (1986) Resilience of animal and plant communities to human disturbance. In *Resilience in Mediterranean-type Ecosystems*, ed. Dell, B., Hopkins, A. J. M. & Lamont, B. B., pp. 39–64. Dordrecht: Junk.

Fox, B. J. & McKay, G. M. (1981) Small mammal responses to pyric successional changes in eucalypt forest. *Australian Journal of Ecology* **6**: 29–41.

Fox, B. J., Fox, M. D. & McKay, G. M. (1979) Litter accumulation after fire in a eucalypt forest. *Australian Journal of Botany* **27**: 157–165.

Fox, B. J., Quinn, R. D. & Breytenbach, J. (1985) A comparison of small mammal succession following fire in shrublands of Australia, California and South Africa. In *Are Australian Ecosystems Different? Proceedings of the Ecological Society of Australia* **14**: 179–197.

Fox, M. D. & Fox, B. J. (1986) The effect of fire frequency on the structure and floristic composition of a woodland understorey. *Australian Journal of Ecology* **11**: 77–85.

Fox, M. D. & Fox, B. J. (1987) The role of fire in the scleromorphic forests and shrublands of eastern Australia. In *The Role of Fire in Ecological Systems*, ed. Trabaud, L., pp. 23–48. Hague: SPB Academic.

Friend, G. R. (1993) Impact of fire on small vertebrates in mallee woodlands and heathlands of temperate Australia: a review. *Biological Conservation* **65**: 99–114.

Friend, G. R. (1994) Post-fire response patterns of invertebrates: are they predictable? *Department of Conservation and Land Management, Western Australia.*

Frost, P. G. H. (1984) The responses and survival of organisms in fire-prone environments. In *Ecological Effects of Fire in South African Ecosystems*, ed. Booysen, P. de V. & Tainton, N. M., pp. 274–309. Berlin: Springer-Verlag.

Fuller, M. (1991) *Forest Fires: An Introduction to Wildland Fire Behavior, Management, Firefighting, and Prevention*. New York: John Wiley and Sons.

Fulton, R. E. & Carpenter, F. L. (1979) Pollination, reproduction and fire in California Arctostaphylos. *Oecologia* **38**: 147–157.

Fyles, J. W. (1989) Seed bank populations in upland coniferous forests in central Alberta. *Canadian Journal of Botany* **67**: 274–278.

Gandar, M. V. (1982) Description of a fire and its effects in the Nylsvley Nature Reserve: a synthesis report. *South African National Science Report Series* **63**: 1–39.

Gatsuk, L. E., Smirnova, O. V., Vorontzova, L. I., Zaugolnova, L. B. & Zhukova, L. A. (1980) Age states of plants of various growth forms: a review. *Journal of Ecology* **68**: 675–696.

Gauthier, S., Bergeron, Y. & Simon, J. (1993) Cone serotiny in jack pine: ontogenic, positional, and environmental effects. *Canadian Journal of Forest Research* **23**: 394–401.

Gee, J. H. R. & Giller, P. S. (1987) *Organization of Communities: Past and Present*. Oxford: Blackwell Scientific.

George, A. S. (1981) The genus Banksia L.f. (Proteaceae). *Nuytsia* **3**: 239–473.

George, A. S. (1984) *The Banksia Book*. Sydney: Kangaroo Press.

Gilbert, J. M. (1959) Forest succession in the Florentine Valley, Tasmania. *Papers and Proceedings of the Royal Society of Tasmania* **93**: 129–151.

Gill, A. M. (1974) Towards an understanding of fire scar formation: field observation and laboratory simulation. *Forest Science* **20**: 198–205.

Gill, A. M. (1975) Fire and the Australian flora: a review. *Australian Forestry* **38**: 4–25.

Gill, A. M. (1976) Fire and the opening of Banksia ornata F. Muell. follicles. *Australian Journal of Botany* **24**: 329–335.

Gill, A. M. (1980) Restoration of bark thickness after fire and mechanical injury in a smooth-barked eucalypt. *Australian Forest Research* **10**: 311–319.

Gill, A. M. (1981a) Coping with fire. In *The Biology of Australian Plants*, ed. Pate, J. S. & McComb, A. J., pp. 65–87. Nedlands, Western Australia: Universtiy of Western Australia Press.

Gill, A. M. (1981b) Adaptive responses of Australian vascular plant species to fires. In *Fire and the Australian Biota*, ed. Gill, A. M., Groves, R. H. & Noble, I., pp. 243–272. Canberra: Australian Academy of Science.

Gill, A. M. (1981c) Fire adaptive traits of vascular plants. In *Fire Regimes and Ecosystem Properties*, ed. Mooney, H. A., Bonnicksen, T. M., Christensen, N. L., Lotan, J. E. & Reiners, W. A., pp. 208–230. USDA Forest Service General Technical Report WO-26.

Gill, A. M. (1981d) Post-settlement fire history in Victorian landscapes. In *Fire and the Australian Biota*, ed. Gill, A. M., Groves, R. H. & Noble, I. R., pp. 77–98. Canberra: Australian Academy of Science.

Gill, A. M. (1983) *Research for the Fire Management of Western Australian State Forests and Conservation Reserves*. Western Australian Department of Conservation and Land Management, Technical Report 12, 82pp.

Gill, A. M. & Ashton, D. H. (1968) The role of bark type in relative tolerance to fire of three central Victorian eucalypts. *Australian Journal of Botany* 16: 491–498.

Gill, A. M. & Bradstock, R. A. (1992) A national register for the fire responses of plant species. *Cunninghamia* 2: 653–660.

Gill, A. M. & Ingwersen, F. (1976) Growth of *Xanthorrhoea australis* R. Br. in relation to fire. *Journal of Applied Ecology* 13: 195–203.

Gill, A. M. & McMahon, A. (1986) A post-fire chronosequence of cone, follicle and seed production in *Banksia ornata*. *Australian Journal of Botany* 34: 425–433.

Gill, A. M., Groves, R. H. & Noble, I. R. (eds.) (1981) *Fire and the Australian Biota*. Canberra: Australian Academy of Science.

Gillon, D. (1972) The effect of bush fire on the principal Pentatomid bugs (Hemiptera) of an Ivory Coast savanna. *Proceedings of the Tall Timbers Fire Ecology Conference* 11: 377–417.

Gillon, Y. (1972) The effect of bushfire on the principal Acridid species of an Ivory Coast savanna. *Proceedings of the Tall Timbers Fire Ecology Conference* 11: 419–471.

Gimingham, C. H., Hobbs, R. J. & Mallik, A. U. (1981) Community dynamics in relation to management of heathland vegetation in Scotland. *Vegetatio* 46: 149–155.

Givnish, T. J. (1981) Serotiny, geography and fire in the Pine Barrens of New Jersey. *Evolution* 35: 101–123.

Glasby, P., Selkirk, P. M., Adamson, D., Downing, A. J. & Selkirk, D. R. (1988) Blue Mountains Ash (*Eucalyptus oreades* R. T. Baker) in the western Blue Mountains. *Proceedings of the Linnean Society of New South Wales* 110: 141–158.

Gleadow, R. M. & Ashton, D. H. (1981) Invasion by *Pittosporum undulatum* of the forests of central Victoria. I. Invasion patterns and plant morphology. *Australian Journal of Botany* 29: 705–720.

Goldammer, J. G. (ed.) (1990) *Fire in the Tropical Biota: Ecosystem Processes and Global Challenges*. Berlin: Springer-Verlag.

Good, R. (1981) Adaptations of Australian plants to fires. In *Bushfires: Their Effect on Australian Life and Landscape*, ed. Stanbury, P., pp. 49–59. Sydney: Macleay Museum, University of Sydney.

Gould, S. J. & Lewontin, R. C. (1979) The spandrels of San Marco and the Panglossian paradigm: a critique of the adaptationist programme. *Proceedings of the Royal Society of London* B205: 581–598.

Granger, J. E. (1984) Fire in forest. In *Ecological Effects of Fire in South African Ecosystems*, ed. Booysen, P. de V. & Tainton, N. M., pp. 177–197. Berlin: Springer-Verlag.

Granström, A. (1987) Seed viability of fourteen species during five years of storage in a forest soil. *Journal of Ecology* 75: 321–331.

Gray, A. J. (1987) Genetic change during succession in plants. In *Colonization, Succession and Stability*, ed. Gray, A. J., Crawley, M. J. & Edwards, P. J., pp. 275–293. Oxford: Blackwell Scientific.

Green, R. H. (1979) *Sampling Design and Statistical Methods for Environmental Biologists*. New York: Wiley.

Greenslade, P & Rosser, G. (1984) Fire and soil-surface insects in the Mount Lofty Ranges, South Australia. In *Medecos IV*, ed. Dell, B., pp. 63–64. Western Australia: Botany Department, University of Western Australia.

Grime, J. P. (1985) Towards a functional description of vegetation. In *The*

Population Structure of Vegetation, ed. White, J., pp. 503–514. Dordrecht: Junk.

Grimm, E. C. (1984) Fire and other factors controlling the Big Woods vegetation of Minnesota in the mid-nineteenth century. *Ecological Monographs* **54**: 291–311.

Grubb, P. J. (1977) The maintenance of species-richness in plant communities: the importance of the regeneration niche. *Biological Reviews* **52**: 107–145.

Grubb, P. J. (1985) Plant populations and vegetation in relation to habitat, disturbance and competition: problems of generalization. In *The Population Structure of Vegetation,* ed. White, J., pp. 595–621. Dordrecht: Junk.

Grubb, P. J. & Hopkins, A. J. M. (1986) Resilience at the level of the plant community. In *Resilience in Mediterranean-type Ecosystems,* ed. Dell, B., Hopkins, A. J. M. & Lamont, B. B., pp. 21–38. Dordrecht: Junk.

Grubb, P. J., Kelly, D. & Mitchley, J. (1982) The control of relative abundance in communities of herbaceous plants. In *The Plant Community as a Working Mechanism,* ed. Newman, E. I., pp. 79–97. Oxford: Blackwell Scientific.

Guthrie, R. D. (1967) Fire melanisms among mammals. *American Midland Naturalist* **77**: 227–230.

Hall, M. (1984) Man's historical and traditional use of fire in southern Africa. In *Ecological Effects of Fire in South African Ecosystems,* ed. Booysen, P. de V. & Tainton, N. M., pp. 40–52. Berlin: Springer-Verlag.

Hallam, S. (1975) *Fire and Hearth.* Australian Institute of Aboriginal Studies, Canberra, Australia.

Handley, C. O. (1969) Fire and mammals. *Proceedings of the Tall Timbers Fire Ecology Conference* **9**: 151–159.

Hanes, T. L. (1971) Succession after fire in the chaparral of southern California. *Ecological Monographs* **41**: 27–52.

Hanes, T. L. (1977) California chaparral. In *Terrestrial Vegetation of California,* ed. Barbour, M. G. & Major, J., pp. 417–469. New York: Wiley.

Hansen, J. D. (1986) Comparison of insects from burned and unburned areas after a range fire. *Great Basin Naturalist* **46**: 721–727.

Hare, R. C. (1961) Heat effects on living plants. *USDA Forest Service Southern Forest Experimental Station Occasional Paper 183.*

Hare, R. C. (1965) Contribution of bark to fire resistance of southern trees. *Journal of Forestry* **63**: 248–251.

Harley, J. L. & Smith, S. E. (1983) *Mycorrhizal Symbiosis.* London: Academic Press.

Harper, J. L. (1977) *The Population Biology of Plants.* London: Academic Press.

Harper, J. L. (1982) After description. In *The Plant Community As A Working Mechanism,* ed. Newman, E. I., pp. 11–25. Oxford: Blackwell Scientific.

Harper, J. L. & White, J. (1971) The dynamics of plant populations. *Proceedings of the Advanced Study Institute of* Dynamics Numbers and Populations (Oosterbeek, 1970) 41–63.

Haskins, M. F. and Shaddy, J. H. (1986) The ecological effects of burning, mowing and plowing on ground-inhabiting spiders (Araneae) in an old-field ecosystem. *Journal of Arachnology* **14**: 1–13.

Hassan, M. A. & West, N. E. (1986) Dynamics of soil seed pools in burned and unburned sagebrush semi-deserts. *Ecology* **67**: 269–272.

Hauge, E. and Kvamme, T. (1983) Spiders from forest-fire areas in southeast Norway. *Fauna Norv. (Series B)* **30**: 39–45.

Hemsley, J. (1967) *Bushfire – S. E. Tasmania*. National Parks and Wildlife Service, Tasmania.

Hendrix, S. D., Brown, V. K. & Gange, A. C. (1988) Effects of insect herbivory on early plant succession: comparison of an English site and an American site. *Biological Journal of the Linnean Society* **35**: 205–216.

Henry, N. B. (1961) Complete protection versus prescribed burning in the Maryborough hardwoods. *Queensland Forest Service Research Note No. 13.*

Hermann, S., Landers, G. and Myers, R. L. (eds.) (1991) *High Intensity Fire in Wildlands: Management Challenges and Options. Proceedings of the Tall Timbers Fire Ecology Conference* **17**.

Heyward, F. (1938) Soil temperatures during forest fires in the Longleaf Pine region. *Journal of Forestry* **36**: 478–491.

Heyward, F. & Tissot, A. N. (1936) Some changes in the soil fauna associated with forest fires in the longleaf pine region. *Ecology* **17**: 659–666.

Hickman, J. C. (1979) The basic biology of plant numbers. In *Topics In Plant Population Biology*, ed. Solbrig, O. T., Jain, S. K. & Johnson, G. B., pp. 232–263. New York: Columbia University Press.

Hilmon, J. B. & Lewis, C. E. (1962) Effect of burning on South Florida range. *USDA Forest Service, Southeastern Forest Experimental Station Paper 146.*

Hils, M. H. & Vankat, J. L. (1982) Species removals from a first-year old-field plant community. *Ecology* **63**: 705–711.

Hindmarsh, R. & Majer, J. D. (1977) Food requirements of the Mardo (*Antechinus flavipes* Waterhouse) and the effect of fire on Mardo abundance. *Forests Department of Western Australia Research Paper 31.*

Hobbs, N. T. & Spowart, R. A. (1984) Effects of prescribed fire on nutrition of mountain sheep and mule deer during Winter and Spring. *Journal of Wildlife Management* **48**: 551–560.

Hobbs, R. J. & Atkins, L. (1990) Fire-related dynamics of a *Banksia* woodland in south-western Western Australia. *Australian Journal of Botany* **38**: 97–110.

Hobbs, R. J. & Gimingham, C. H. (1984) Studies of fire in Scottish heathland communities. I. Fire characteristics. *Journal of Ecology* **72**: 223–240.

Hobbs, R. J. & Gimingham, C. H. (1987) Vegetation, fire and herbivore interactions in heathland. *Advances in Ecological Research* **16**: 87–159.

Hobbs, R. J., Mallik, A. U. & Gimingham, C. H. (1984) Studies on fire in Scottish heathland communities. III. Vital attributes of the species. *Journal of Ecology* **72**: 963–976.

Hocking, B. (1964) Fire melanism in some African grasshoppers. *Evolution* **18**: 332–335.

Hodgkins, E. J. (1958) Effects of fire on undergrowth vegetation in upland southern pine forests. *Ecology* **39**: 36–46.

Holland, P. G. (1986) Mallee vegetation: steady state or successional? *Australian Geographer* **17**: 113–120.

Hopkins, A. J. M. & Griffin, E. A. (1984) Floristic patterns. In *Kwongan: Plant Life on the Sandplain*, ed. Pate, J. S. & Beard, J. S., pp. 69–83. Western Australia: University of Western Australia Press.

Hopper, S. D. (1979) Biogeographical aspects of speciation in the southwest Australian flora. *Annual Review of Ecology and Systematics* **10**: 399–422.

Houston, D. B. (1973) Wildfires in northern Yellowstone National Park. *Ecology* **54**: 1111–1117.

Howard, T. M. (1973) Studies in the ecology of *Nothofagus cunninghamii* Oerst. I. Natural regeneration on the Mt. Donna Buang massif, Victoria. *Australian Journal of Botany* **21**: 67–78.

Howard, W. E., Fenner, R. L. & Childs, H. E. (1959) Wildlife survival in brush burns. *Journal of Rangeland Management* **12**: 230–234.

Hughes, L. & Westoby, M. (1990) Removal rates of seeds adapted for dispersal by ants. *Ecology* **71**: 138–148.

Hughes, L. & Westoby, M. (1992) Fate of seeds adapted for dispersal by ants in Australian sclerophyll vegetation. *Ecology* **73**: 1285–1299.

Hurlbert, S. H. (1984) Pseudoreplication and the design of ecological field experiments. *Ecological Monographs* **54**: 187–211.

Ingwersen, F. (1977) Vegetation Development after Fire in the Jervis Bay Territory. Unpublished M. Sc. Thesis, Australian National University.

Ito, M. & Iizumi, S. (1960) Temperatures during grassland fires and their effect on some species in Kawatabi, Miyagi Prefecture. *Tokyo Uni. Sci. Rep. Res. Inst.* D-11: 109–114.

Izzara, de D. C. (1977) The effects of buring on soil micro-arthropods in the semi-arid pampa zone. In *Soil Organisms as Components of Ecosystems*, ed. Lohm, U. & Persson, T., *Ecological Bulletin* **25**: 357–365.

Jackson, W. D. (1968) Fire, air, water and earth: an elemental ecology of Tasmania. *Proceedings of the Ecological Society of Australia* **3**: 9–16.

Jacobs, M. R. (1955) *The Growth Habits of the Eucalypts*. Canberra: Commonwealth of Australia, Forest and Timber Bureau.

James, S. (1984) Lignotubers and burls: their structure, function and ecological significance in Mediterranean ecosystems. *Botanical Reviews* **50**: 225–266.

Jarrett, P. H. & Petrie, A. H. K. (1929) The vegetation of the Blacks' Spur region. A study in the ecology of some Australian mountain *Eucalyptus* forests. II. Pyric succession. *Journal of Ecology* **17**: 249–281.

Jocque, R. (1981) On reduced size in spiders from marginal habitats. *Oecologia* **49**: 404–408.

Johnson, E. A. (1975) Buried seed populations in the subarctic forest east of Great Slav Lake, Northwest Territories. *Canadian Journal of Botany* **53**: 2933–2941.

Johnson, E. A. (1992) *Fire and Vegetation Dynamics: Studies from the North American Boreal Forest*. Cambridge: Cambridge University Press.

Jones, J. S. (1982) More to melanism than meets the eye. *Nature* **300**: 109–110.

Jones, R. (1969) Fire-stick farming. *Australian Natural History* **September**: 224–228.

Jordaan, P. G. (1949) Aantekeninge oor die voortplanting en brandperiodes van *Protea mellifera* Thunb. *Journal of South African Botany* **15**: 121–125.

Jordaan, P. G. (1965) Die invloed van 'n winterbrand op die voortplanting van vier soorte van die Proteaceae. *Tydskrif vir Natuurbewaaring* **5**: 27–31.

Jordan, R. (1984) Barren Grounds Bird Observatory report 1982–84. *RAOU Report No. 11* pp. 56–57.

Kalisz, P. J. (1982) The longleaf pine islands of the Ocala National Forest, Florida: a soil study. PhD Dissertation, University of Florida.

Kauffman, J. B. & Uhl, C. (1990) Interactions of anthropogenic activities, fire and rain forests in the Amazon Basin. In *Fire in the Tropical Biota: Ecosystem Processes and Global Challenges*, ed. Goldammer, J. G., pp. 117–134. Berlin: Springer-Verlag.

Keeley, J. E. (1977) Seed production, seed populations in soil and seedling production after fire for two congeneric pairs of sprouting and non-sprouting chaparral shrubs. *Ecology* **58**: 820–829.

Keeley, J. E. (1981) Reproductive cycles and fire regimes. In *Fire Regimes and Ecosystem Properties*, pp. 231–277. *USDA Forest Service General Technical Report WO-26*.

Keeley, J. E. (1986) Seed germination patterns of *Salvia mellifera* in fire-prone environments. *Oecologia* **71**: 1–5.

Keeley, J. E. (1992) Recruitment of seedlings and vegetative sprouts in unburned chaparral. *Ecology* **73**: 1194–1208.

Keeley, J. E. & Keeley, S. C. (1984) Postfire recovery of California coastal sage scrub. *American Midland Naturalist* **111**: 105–117.

Keeley, J. E. & Zedler, P. H. (1978) Reproduction of chaparral shrubs after fire: a comparison of sprouting and seeding strategies. *American Midland Naturalist* **99**: 142–161.

Keeley, J. E., Morton, B. A., Pedrosa, A. & Trotter, P. (1985) Role of allelopathy, heat and charred wood in the germination of chaparral herbs and suffrutescents. *Journal of Ecology* **73**: 445–458.

Keeley, S. C. & Keeley, J. E. (1982) The role of allelopathy, heat, and charred wood in the germination of chaparral herbs. In *Proceedings of a Symposium on Dynamics and Management of Mediterranean Type Ecosystems*, ed. Conrad, C. E. & Oechel, W. C., pp. 128–134. *USDA Forest Service General Technical Report PSW-58*.

Keith, D. A. (1991) Coexistence and species diversity in upland swamp vegetation: the roles of an environmental gradient and recurring fire. Unpublished PhD Thesis, University of Sydney.

Keith, L. B. & Surrendi, D. C. (1971) Effects of fire on a snowshoe hare population. *Journal of Wildlife Management* **35**: 16–26.

Kern, N. G. (1978) The influence of fire on populations of small mammals of the Kruger National Park. Unpublished MSc Thesis, University of Pretoria.

Kessell, S. R. & Cattelino, P. J. (1978) Evaluation of a fire behaviour information integration system for southern California chaparral wildlands. *Environmental Management* **2**: 135–159.

Kettlewell, B. (1973) *The Evolution of Melanisms*. Oxford: Clarendon Press.

Kikkawa, J. & Anderson, D. J. (1986) *Community Ecology: Pattern and Process*. Melbourne: Blackwell Scientific.

Kikkawa, J., Ingram, G. J. & Dwyer, P. D. (1979) The vertebrate fauna of Australian heathlands: an evolutionary perspective. In *Heathlands and Related Shrublands of the World. 9A. Descriptive Studies*, ed. Specht, R. L., pp. 231–279. Amsterdam: Elsevier.

Kikkawa, J., Monteith, G. B. & Ingram, G. (1981) Cape York Peninsular: major region of faunal interchange. In *Ecological Biogeography of Australia*, ed. Keast, A., pp. 1695–1742. The Hague: Junk.

Kilgore, B. M. (1973) The ecological role of fire in Sierran conifer forests: its application to National Park management. *Quaternary Research* **3**: 496–513.

Kiltie, R. A. (1989) Wildfire and the evolution of dorsal melanism in fox squirrels, *Sciurus niger*. *Journal of Mammalogy* **70**: 726–739.

Kimber, P. C. (1978) Increased girth increment associated with crown scorch in Jarrah. *Forests Department of Western Australia Research Paper 37*.

King, N. K. & Vines, R. G. (1969) Variation in the flammability of the leaves of some Australian forest species. *CSIRO Mimeograph Report, Division of Applied Chemistry, Melbourne*.

King, T. J. & Woodell, S. R. J. (1973) The causes of regular pattern in desert perennials. *Journal of Ecology* **61**: 761–765.

Klopatek, C. C., Debano, L. F. & Klopatek, J. M. (1988) Effects of simulated fire on vesicular–arbuscular mycorrhizae in pinyon–juniper woodland soil. *Plant and Soil* **109**: 245–249.

Knapp, A. K. (1984) Post-burn differences in solar radiation, leaf temperature and water stress influencing production in a lowland tallgrass prairie. *American Journal of Botany* **71**: 220–227.

Koch, J. M. & Bell, D. T. (1980) Leaf scorch in *Xanthorrhoea gracilis* as an index of fire intensity. *Australian Forest Research* **10**: 113–119.

Koch, L. E. & Majer, J. D. (1980) A phenological investigation of various invertebrates in forest and woodland areas in the south-west of Western Australia. *Journal of the Royal Society of Western Australia* **63**: 21–28.

Komarek, E. V. (1964) The natural history of lightning. *Proceedings of the Tall Timbers Fire Ecology Conference* **3**: 139–183.

Komarek, E. V. (1965) Fire ecology – grasslands and Man. *Proceedings of the Tall Timbers Fire Ecology Conference* **4**: 169–220.

Komarek, E. V. (1967) Fire – and the ecology of Man. *Proceedings of the Tall Timbers Fire Ecology Conference* **6**: 143–170.

Komarek, E. V. (1968) The nature of lightning fires. *Proceedings of the Tall Timbers Fire Ecology Conference* **7**: 5–41.

Komarek, E. V. (1969) Lightning and lightning fires as ecological forces. *Proceedings of the Tall Timbers Fire Ecology Conference* **8**: 169–197.

Krebs, C. J. (1985) *Ecology: The Experimental Study of Distribution and Abundance*. 3rd edn., New York: Harper and Row.

Kruger, F. J. (1977) Ecology of Cape fynbos in relation to fire. In *Proceedings of a Symposium on the Environmental Consequences of Fire and Fuel Management in Mediterranean Ecosystems*, ed. Mooney, H. A. & Conrad, C. E., pp. 230–244. *USDA Forest Service General Technical Report WO-3*.

Kruger, F. J. (1983) Plant community diversity and dynamics in relation to fire. In *Mediterranean-Type Ecosystems: The Role of Nutrients*, ed. Kruger, F. J., Mitchell, D. T. and Jarvis, J. U. M., pp. 446–472. Berlin: Springer-Verlag.

Kruger, F. J. & Bigalke, R. C. (1984) Fire in fynbos. In *Ecological Effects of Fire in South African Ecosystems*, ed. Booysen, P. de V. and Tainton, N. M., pp. 67–114. Berlin: Springer-Verlag.

Kucera, C. L. & Ehrenreich, J. H. (1962) Some effects of annual burning on central Missouri prairie. *Ecology* **43**: 334–336.

Laessle, A. M. (1942) *The Plant Communities of the Welaka Area with Special Reference to Correlation Between Soils and Vegetational Succession*. Biological Science Series **4**. Gainesville: University of Florida Publications.

Lamont, B. B. (1985) The comparative reproductive biology of three *Leucospermum* species (Proteaceae) in relation to fire responses and breeding system. *Australian Journal of Botany* **33**: 139–145.

Lamont, B. B. (1991) Canopy seed storage and release – what's in a name? *Oikos* **60**: 1–3.

Lamont, B. B. (1993) Injury-induced cyanogenesis in vegetative and reproductive parts of two *Grevillea* species and their F₁ hybrid. *Annals of Botany* **71**: 537–542.

Lamont, B. B. & Barker, M. J. (1988) Seed bank dynamics of a serotinous, fire-sensitive *Banksia* species. *Australian Journal of Botany* **36**: 193–203.

Lamont, B. B. & Runciman, H. V. (1993) Fire may stimulate flowering, branching, seed production and seedling establishment in two kangaroo paws (Haemodoraceae). *Journal of Applied Ecology* **30**: 256–264.

Lamont, B. B., Collins, B. C. & Cowling, R. M. (1985) Reproductive biology of the Proteaceae in Australia and South Africa. *Proceedings of the Ecological Society of Australia* **14**: 213–224.

Lamont, B. B., Connell, S. W. & Bergl, S. M. (1991a) Seed bank and population dynamics of *Banksia cuneata*: the role of time, fire and moisture. *Botanical Gazette* **152**: 114–122.

Lamont, B. B., Le Maitre, D. C., Cowling, R. M. & Enright, N. J. (1991b) Canopy seed storage in woody plants. *Botanical Reviews* **57**: 277–317.

Lamont, B. B., Witkowski, E. T. F. & Enright, N. J. (1993) Post-fire litter microsites: safe for seeds, unsafe for seedlings. *Ecology* **74**: 501–512.

Lance, A. N. (1983) Performance of sheep on unburned and serially burned blanket bog in western Ireland. *Journal of Applied Ecology* **20**: 767–775.

Langdon, O. G. (1981) Some effects of prescribed fire on understory vegetation in loblolly pine stands. In *Prescribed Fire and Wildlife in Southern Forests: Proceedings of a Symposium,* ed. Wood, G. W., pp. 143–153. Georgetown, South Carolina: The Belle W. Baruch Forest Science Institute, Clemson University.

Law, R. (1979) The cost of reproduction in annual meadow grass. *American Naturalist* **113**: 3–16.

Lawrence, G. E. (1966) Ecology of vertebrate animals in relation to chaparral fire in the Sierra Nevada foothills. *Ecology* **47**: 278–291.

Le Houerou, H. N. (1974) Fire and vegetation in the Mediterranean Basin. *Proceedings of the Tall Timbers Fire Ecology Conference* **13**: 237–277.

Le Maitre, D. C. (1985) Current interpretations of the term serotiny. *South African Journal of Science* **81**: 289–290.

Le Maitre, D. C. (1987) Effects of season of burn on species populations and composition of fynbos in the Jonkershoek valley. *South African Journal of Botany* **53**: 284–292.

Le Maitre, D. C. (1992) The relative advantages of seeding and sprouting in fire-prone environments: a comparison of life histories of *Protea neriifolia* and *Protea nitida*. In *Fire in South African Mountain Fynbos: Ecosystem, Community and Species Response at Swartboskloof,* ed. van Wilgen, B. W., Richardson, D. M., Kruger, F. J. & van Hensbergen, H. J., pp. 123–144. Berlin: Springer-Verlag.

Le Maitre, D. C. & Brown, P. J. (1992) Life cycles and fire-stimulated flowering in geophytes. In *Fire in South African Mountain Fynbos: Ecosystem, Community and Species Response at Swartboskloof,* ed. van Wilgen, B. W., Richardson, D. M., Kruger, F. J. & van Hensbergen, H. J., pp. 145–160. Berlin: Springer-Verlag.

Leigh, J. H. & Holgate, M. D. (1979) Responses of understorey of forests and woodlands of the southern tablelands to grazing and burning. *Australian Journal of Ecology* **4**: 25–45.

Leonard, B. V. (1974) The effects of burning on litter fauna in eucalypt forests. *Proceedings of the 3rd Fire Ecology Symposium, Monash University*, pp. 43–48.

Leverich, W. J. & Levin, D. A. (1979) Age-specific survivorship and reproduction in Phlox drummondii. *American Naturalist* **113**: 881–903.

Levitt, J. (1972) *Responses of Plants to Environmental Stresses*. New York: Academic Press.

Lilly, C. E. and Hobbs, G. A. (1962) Effects of spring burning and insecticides on the Superb Plant Bug, *Adelphocoris superbus* (Uhl.) and associated fauna in alfalfa seed fields. *Canadian Journal of Plant Science* **42**: 53–61.

Lillywhite, H. B. (1977) Animal responses to fire and fuel management in chaparral. In *Proceedings of a Symposium on the Environmental Consequences of Fire and Fuel Management in Mediterranean Ecosystems*, ed. Mooney, H. A. & Conrad, C. E., pp. 368–372. *USDA Forest Service General Technical Report WO-3*.

Lillywhite, H. B., Friedman, G. & Ford, N. (1977) Color matching and perch selection by lizards in recently burned chaparral. *Copeia* **1**: 115–121.

Lindemuth, A. W. & Davis, J. R. (1973) Predicting fire spread in Arizona's oak chaparral. *USDA Forest Service Research Papers RM-101*.

Linhart, Y. B. (1978) Maintenance of variation in cone morphology in California closed-cone pines: the roles of fire, squirrels and seed output. *Southwestern Naturalist* **23**: 29–40.

Little, S. (1974) Effects of fire on temperate forests: north-eastern United States. In *Fire and Ecosystems*, ed. Kozlowski, T. T. & Ahlgren, C. E., pp. 225–250. New York: Academic Press.

Lloret, F. & Zedler, P. H. (1991) Recruitment pattern of *Rhus integrifolia* populations in periods between fire in chapparral. *Journal of Vegetation Science* **2**: 217–230.

Lloyd, M. (1967) Meaning crowding. *Journal of Animal Ecology* **36**: 1–30.

Lonsdale, W. M. & Braithwaite, R. W. (1991) Assessing the effects of fire on vegetation in tropical savannas. *Australian Journal of Ecology* **16**: 363–374.

Lopez-Soria, L. & Castell, C. (1992) Comparative genet survival after fire in woody Mediterranean species. *Oecologia* **91**: 493–499.

Lorimer, C. G. (1980) The use of land survey records in estimating pre-settlement fire frequency. In *Proceedings of the Fire History Workshop*, ed. Stokes, M. A. and Dieterich, J. H., pp. 57–62. *USDA Forest Service General Technical Report RM-81*.

Lotan, J. E. (1976) Cone serotiny – fire relationships in Lodgepole Pine. *Proceedings of the Tall Timbers Fire Ecology Conference* **14**: 267–278.

Lotti, T., Klawitter, R. A. & LeGrande, W. P. (1960) Prescribed buring for understory control in loblolly pine stands of the coastal plain. *USDA Forest Service Southeast Forest Experimental Station Paper 116*.

Lovat, L. (1911) Heather-burning. In *The Grouse in Health and Disease, Being the Final Report of the Committee of Inquiry on Grouse Disease*, ed. Leslie, A. S., pp. 392–413. London: Smith, Elder & Co.

Luke, R. H. & McArthur, A. G. (1978) *Bushfires In Australia*. Canberra: Australian Government Publishing Service.

Lussenhop, J. (1976) Soil arthropod response to prairie burning. *Ecology* **57**: 88–98.

Lutz, H. J. (1956) Ecological effects of forest fires in the interior of Alaska. *Technical Bulletin No. 1133, US Department of Agriculture*.

Lutz, H. J. (1960) Fire as an ecological factor in the boreal forest of Alaska. *Journal of Forestry* **58**: 454–460.

Main, A. R. (1981) Fire tolerance of heathland animals. In *Heathlands and Related Shrublands of the World. 9B. Analytical Studies*, ed. Specht, R. L., pp. 85–90. Amsterdam: Elsevier.

Majer, J. D. (1980) The influence of ants on broadcast and naturally spread seeds in rehabilitated bauxite mined areas. *Reclamation Review* **3**: 3–9.

Majer, J. D. (1984) Short term responses of soil and litter invertebrates to a cool autumn burn in Jarrah (*Eucalyptus marginata*) forest in Western Australia. *Pedobiologia* **26**: 229–247.

Malajczuk, N., Trappe, J. M. & Molina, R. (1987) Interrelationships among some ectomycorrhizal trees, hypogeous fungi and small mammals: western Australian and northwestern American parallels. *Australian Journal of Ecology* **12**: 53–55.

Mallik, A. U. (1986) Near-ground micro-climate of burned and unburned *Calluna* heathland. *Journal of Environmental Management* **23**: 157–171.

Mallik, A. U., Hobbs, R. J. & Legg, C. J. (1984b) Seed dynamics in *Calluna–Arctostaphylos* heath in north-eastern Scotland. *Journal of Ecology* **72**: 855–871.

Martin, R. E. & Cushwa, C. T. (1966) Effects of heat and moisture on leguminous seed. *Proceedings of the Tall Timbers Fire Ecology Conference* **5**: 159–175.

McAnninch, J. B. & Strayer, D. L. (1989) What are the tradeoffs between the immediacy of management needs and the longer process of scientific discovery? In *Long-term Studies in Ecology: Approaches and Alternatives*, ed. Likens, G. E., pp. 203–205. Berlin: Springer-Verlag.

McArthur, A. G. (1966) Weather and grassland fire behaviour. *Commonwealth of Australia Forest and Timber Bureau Leaflet No. 100.*

McArthur, A. G. (1967) Fire behaviour in eucalypt forests. *9th Commonwealth Forestry Conference.* Canberra: Australian Forest and Timber Bureau.

McArthur, A. G. (1970) Introduction. In *2nd Fire Ecology Symposium*, pp. 1–22. Australia: Monash University.

McArthur, A. G. (1962) Control burning in eucalypt forests. *8th Commonwealth Forestry Conference*, Australian Forest and Timber Bureau.

McArthur, A. G. & Cheney, N. P. (1966) The characterization of fires in relation to ecological studies. *Australian Forest Research* **2**: 36–45.

McArthur, A. G. & Cheney, N. P. (1972) Source notes on forest fire control. Unpublished Report. Canberra: Forest Research Institute, Australian Forest & Timber Bureau.

McCormick, I. (1968) Succession. *Via* **1**:1–16.

McKeon, G. M. & Mott, J. J. (1982) The effects of temperature on the field softening of hard seed of *Stylosanthes humilis* and *S. hamata* in a dry monsoonal climate. *Australian Journal of Agricultural Research* **33**: 75–85.

McLoughlin, S. P. & Bowers, J. E. (1982) Effects of wildfire on a Sonoran Desert plant community. *Ecology* **63**: 246–248.

McNamara, P. (1955) A preliminary investigation of the fauna of humus layers in the jarrah forest in Western Australia. *Commonwealth of Australia Forest and Timber Bureau Leaflet No. 71.*

McPherson, J. K. & Muller, C. H. (1969) Allelopathic effect of *Adenostoma fasciculatum* chamise, in the California chaparral. *Ecological Monographs* **39**: 177–198.

Mentis, M. T. & Rowe-Rowe, D. T. (1979) Fire and faunal abundance and diversity in the Natal Drakensberg. *Proceedings of the Grassland Society of South Africa* **14**: 75–77.

Meredith, C. W., Gilmore, A. M. & Isles, A. C. (1984) The ground parrot (*Pezoporus wallicus* Kerr) in south-eastern Australia: a fire-adapted species? *Australian Journal of Ecology* **9**: 367–380.

Merrett, P. (1976) Changes in the ground-living spider fauna after heathland fires in Dorset. *Bulletin of the British Arachnology Society* **3**: 214–221.

Merrilees, D., Dix, W. C., Hallam, S. H., Douglas, W. H. & Berndt, R. M. (1973) Aboriginal man in southwestern Australia. *Proceedings of the Royal Society of Western Australia* **56**: 44–55.

Metz, L. J. & Farrier, M. H. (1971) Prescribed burning and soil mesofauna on the Santee Experimental Forest. In *Proceedings of the Prescribed Burning Symposium*, pp. 100–106. USDA Forest Service Southeast Forest Experimental Station, North Carolina.

Midgley, J. J. (1989) Season of burn of serotinous fynbos Proteacae: a critical review and further data. *South African Journal of Botany* **55**: 165–170.

Midgley, J. J. & Clayton, P. (1990) Short-term effects of an autumn fire on small mammal populations in Southern Cape coastal mountain fynbos. *South African Forestry Journal* **153**: 27–30.

Mills, J. N (1986) Herbivores and early postfire succession in southern California chaparral. *Ecology* **67**: 1637–1649.

Minnich, R. A. (1983) Fire mosaics in southern California and northern Baja California. *Science* **219**: 1287–1294.

Minnich, R. A. (1988) *The Biogeography of Fire in the San Bernadino Mountains of California: A Historical Study*. Berkeley: University of California Press.

Mirov, N. T. (1936) Germination behaviour of some California plants. *Ecology* **17**: 667–672.

Mitchell, T. L. (1848) *Journal of an Expedition into the Interior of Tropical Australia*. London: Brown, Green & Longman.

Mohamed, B. F. & Gimingham, C. H. (1970) The morphology of vegetative regeneration in *Calluna vulgaris*. *New Phytologist* **69**: 743–750.

Mohr, C. (1896) The timber pines of the southern United States. *USDA Division of Forestry Bulletin* **13**: 49–66.

Monk, C. D. (1968) Successional and environmental relationships of the forest vegetation of north central Florida. *American Midland Naturalist* **79**: 441–457.

Mooney, H. A., Bonnickson, T. M., Christensen, N. L., Lotan, J. E. & Reiners, W. A. (eds.) (1981) *Fire Regimes and Ecosystem Properties*. USDA Forest Service General Technical Report WO-26.

Mooney, H. E. & Conrad, C. E. (eds.) (1977) *Environmental Consequences of Fire and Fuel Management in Mediterranean Ecosystems*. USDA Forest Service General Technical Report WO-3.

Moore, J. M. & Wein, R. W. (1977) Viable seed populations by soil depth and potential site recolonization after disturbance. *Canadian Journal of Botany* **55**: 2408–2411.

Moreno, J. M. & Oechel, W. C. (1989) A simple method for estimating fire intensity after a burn in California chaparral. *Acta Oecologia/Oecologia Plantarum* **10**: 57–68.

Moreno, J. M. & Oechel, W. (1991) Fire intensity effects on germination of shrubs and herbs in southern California chaparral. *Ecology* **72**: 1993–2004.

Morris, E. C. & Myerscough, P. J. (1988) Survivorship, growth and self-thinning in *Banksia ericifolia*. *Australian Journal of Ecology* **13**: 181–189.

Mount, A. B. (1979) Natural regeneration processes in Tasmanian forests. *Search* **10**: 180–186.

Muller, C. H. (1965) Inhibitory terpenes volatilised from Salvia shrubs. *Bulletin of the Torrey Botany Club* **93**: 332–351.

Muller, C. H., Muller, W. H. & Haines, B. L. (1964) Volatile growth inhibitors produced by aromatic shrubs. *Science* **143**: 471–473.

Muller, C. H., Hanawalt, R. B. & McPherson, J. K. (1968) Allelopathic control of herb growth in the fire cycle of Californian chaparral. *Bulletin of the Torrey Botany Club* **95**: 225–231.

Muller, W. H. (1965) Volatile materials produced by *Salvia leucophylla*: effects on seedling growth and soil bacteria. *Botanical Gazette* **126**: 195–200.

Munger, T. T. (1940) The cycle from douglas fir to hemlock. *Ecology* **21**: 451–459.

Murie, M. (1963) Homing and orientation of deermice. *Journal of Mammalogy* **44**: 338–349.

Muston, R. M. (1987) A long-term study of fire-induced changes in a Hawkesbury Sandstone plant community. Unpublished Ph. D. Thesis, University of Wollongong, Australia.

Mutch, R. W. (1970) Wildland fires and ecosystems: a hypothesis. *Ecology* **51**: 1046–1051.

Mutch, R. W. (1980) Who cares about fire history? In *Proceedings of the Fire History Workshop*, ed. Stokes, M. A. & Dietrich, J. H., pp. 138–140. USDA Forest Service General Technical Report RM-81.

Myers, R. L. (1985) Fire and the dynamic relationship between Florida sandhill and sand pine scrub vegetation. *Bulletin of the Torrey Botanical Club* **112**: 241–252.

Myers, R. L. (1990) Scrub and high pine. In *Ecosystems of Florida*, ed. Myers, R. L. & Ewel, J. J., pp. 150–193. Orlando: University of Central Florida Press.

Myers, R. L. (1991) Condominiums, trailer parks and high-intensity fires: the future of sand pine scrub preserves in Florida. *Proceedings of the Tall Timbers Fire Ecology Conference* **17**: 301.

Myers, R. L. & Ewel, J. J. (eds.) (1990) *Ecosystems of Florida*. Orlando: University of Central Florida Press.

Myerscough, P. J. (1980) Biological flora of the British Isles: *Epilobium angustifolium* L. *Journal of Ecology* **68**: 1047–1074.

Nakagoshi, N., Nehira, K. & Takahashi, F. (1987) The role of fire in pine forests of Japan. In *The Role of Fire in Ecological Systems*, ed. Trabaud, L., pp. 91–119. Hague: SPB Academic.

Naveh, Z. (1974) The ecology of fire in Israel. *Proceedings of the Tall Timbers Fire Ecology Conference* **13**: 131–170.

Naveh, Z. (1975) The evolutionary significance of fire in the Mediterranean region. *Vegetatio* **29**: 199–208.

Neal, J. L., Wright, E. & Bollen, W. B. (1965) Burning Douglas-fir slash: physical, chemical and microbial effects on the soil. *Forest Research Laboratory Research Paper*. Corvallis: Oregon State University.

Newsome, A. E. (1970) An experimental attempt to produce a mouse plague. *Journal of Animal Ecology* **39**: 299–311.

Newsome, A. E. & Catling, P. C. (1983) Animal demography in relation to fire and shortage of food: some indicative models. In *Mediterranean-type Ecosystems: The Role of Nutrients*, ed. Kruger, F. J., Mitchell, D. T. & Jarvis, J. U. M., pp. 490–505. Berlin: Springer-Verlag.

Newsome, A. E., McIlroy, J. & Catling, P. (1975) The effects of an extensive wildfire on populations of twenty ground vertebrates in south-east Australia. *Proceedings of the Ecological Society of Australia* **9**: 107–123.

Nicholson, P. H. (1981) Fire and the Australian aborigine: an enigma. In *Fire and the Australian Biota*, ed. Gill, A. M., Groves, R. H. & Noble, I. R., pp. 55–76. Canberra: Australian Academy of Science.

Nieuwenhuis, A. (1987) The effect of fire frequency on the sclerophyll vegetation of the West Head, New South Wales. *Australian Journal of Ecology* **12**: 373–385.

Noble, I. R. (1981) Predicting successional change. In *Fire Regimes and Ecosystem Properties*, ed. Mooney, H. A., Bonnicksen, T. M., Christensen, N. L., Lotan, J. E. & Reiners, W. A., pp. 278–300. USDA Forest Service General Technical Report WO-26.

Noble, I. R. & Slatyer, R. O. (1977) Post-fire succession of plants in Mediterranean ecosystems. In *Proceedings of a Symposium on the Environmental Consequences of Fire and Fuel Management in Mediterranean Ecosystems*, pp. 27–36. USDA Forest Service General Technical Report WO-3.

Noble, I. R. & Slatyer, R. O. (1980) The use of vital attributes to predict successional changes in plant communities subject to recurrent disturbance. *Vegetatio* **43**: 5–21.

Noble, I. R., Bary, G. A. V. & Gill, A. M. (1980) McArthur's fire danger meters expressed as equations. *Australian Journal of Ecology* **5**: 201–203.

Noble, J. C. (1982) The significance of fire in the biology and evolutionary ecology of mallee *Eucalyptus* populations. In *Evolution of the Flora and Fauna of Arid Australia*, ed. Barker, W. R. & Greenslade, P. J. M., pp. 153–166. Adelaide: Peacock Publications.

Noble, J. C. (1984) Fire in mallee (*Eucalyptus* spp.) shrublands of southwestern New South Wales. In *Medecos IV*, ed. Dell, B., pp. 130–131. Botany Department, University of Western Australia.

Noble, J. C. (1989) Fire studies in mallee (*Eucalyptus* spp.) communities of western N. S. W.: The effects of fires applied in different seasons on herbage productivity and their implication for management. *Australian Journal of Ecology* **14**: 169–187.

Noble, J. C., Harrington, G. N. & Hodgkinson, K. C. (1985) The ecological significance of irregular fire in Australian rangelands. *Proceedings of the 2nd International Rangeland Congress, Adelaide, Australia*.

Noy-Meir, I. & van der Maarel, E. (1987) Relations between community theory and community analysis in vegetation science: some historical perspectives. *Vegetatio* **69**: 5–15.

O'Dowd, D. J. & Gill, A. M. (1984) Predator satiation and site alteration following fire: mass reproduction of alpine ash (*Eucalyptus delegatensis*) in southeastern Australia. *Ecology* **65**: 1052–1066.

Odum, E. P. (1969) The strategy of ecosystem development. *Science* **164**: 262–270.

Old, S. M. (1969) Microclimate, fire and plant production in an Illinois prairie. *Ecological Monographs* **39**: 355–384.

Oliver, M. D. N., Short, N. R. M. & Hanks, J. (1978) Population ecology of oribi, grey rhebuck and mountain reedbuck in Highmoor State Forest land, Natal. *South African Journal of Wildlife Research* **8**: 95–105.

Olson, J. S. (1958) Rates of succession and soil changes on Southern Lake Michigan sand dunes. *Botanical Gazette* **119**: 125–170.

Packham, D. R. (1970) Heat transfer above a small ground fire. *Australian Forest Research* **5**: 19–24.

Pannell, J. R. & Myerscough, P. J. (1993) Canopy-stored seed banks of *Allocasuarina distyla* and *A. nana* in relation to time since fire. *Australian Journal of Botany* **41**: 1–9.

Pate, J. S. & Beard, J. S. (eds.) (1984) *Kwongan: Plant Life on the Sandplain*. Nedlands, Western Australia: University of Western Australia Press.

Pate, J. S., Dixon, K. W. & Orshan, G. (1984) Growth and life form characteristics of Kwongan species. In *Kwongan: Plant Life on the Sandplain*, ed. Pate, J. S. & Beard, J. S., pp. 84–100. Nedlands, Western Australia: University of Western Australia.

Pen, L. J. & Green, J. W. (1983) Botanical exploration and vegetational changes on Rottnest Island. *Journal of the Royal Society of Western Australia* **66**: 20–24.

Perry, D. A. & Lotan, J. E. (1979) A model of fire selection for serotiny in lodgepole pine. *Evolution* **33**: 958–968.

Phillips, E. P. (1919) A preliminary report on the veld burning experiments at Groenkloof, Pretoria. *South African Journal of Science* **16**: 286–299.

Phillips, E. P. (1920) Veld burning experiments at Groenkloof. Second report. *South African Department of Agriculture Science Bulletin* **17**.

Phillips, J. (1936) Fire in vegetation: a bad master, a good servant and a national problem. *Journal of South African Botany* **January**: 36–45.

Phillips, J. (1965) Fire – as a master and servant: its influence in the bioclimatic regions of trans-Saharan Africa. *Proceedings of the Tall Timbers Fire Ecology Conference* **4**: 7–109.

Pianka, E. (1992) Disturbance, spatial heterogeneity, and biotic diversity: fire succession in arid Australia. *National Geographic Research and Exploration* **8**: 352–371.

Pickett, S. T. A. (1989) Space-for-time substitution as an alternative to long-term studies. In *Long-term Studies In Ecology: Approaches and Alternatives*, ed. Likens, G. E., pp. 110–135. Berlin: Springer-Verlag.

Picozzi, N. (1968) Grouse bags in relation to the management and geology of heather moors. *Journal of Applied Ecology* **5**: 483–487.

Pielou, E. C. (1974) *Population and Community Ecology: Principles and Methods*. New York: Gordon and Breach.

Pierce, S. M. & Cowling, R. M. (1991) Dynamics of soil-stored seed banks of six shrubs in fire-prone dune Fynbos. *Journal of Ecology* **79**: 731–747.

Platt, W. J. & Schwartz, M. (1990) Temperate hardwood forests. In *Ecosystems of Florida*, ed. Myers, R. & Ewel, J. J., pp. 194–229. Orlando: Academic Press.

Platt, W. J., Evans, G. W. & Davis, M. M. (1988) Effects of fire season on flowering of forbs and shrubs in longleaf pine forests. *Oecologia* **76**: 353–363.

Pompe, A. & Vines, R. G. (1966) The influence of moisture on the combustion of leaves. *Australian Forestry* **30**: 231–241.

Posamentier, H. G., Clark, S. S., Hain, D. L. & Recher, H. F. (1981) Succession following wildfire in coastal heathland (Nadgee Nature Reserve N. S. W.). *Australian Journal of Ecology* **6**: 165–175.

Poulton, E. B. (1915) The habits of the Australian Buprestid Fire Beetle, *Merimna atrata*, Lap., et Gory. *Transactions of the Entomological Society of London* **Part 1**: iii–iv.

Preece, N. (1990) Application of fire ecology to fire management: a manager's perspective. *Proceedings of the Ecological Society of Australia* **16**: 221–233.

Press, A. J. (1988) Comparisons of the extent of fire in different land management systems in the Top End of the Northern Territory. *Proceedings of the Ecological Society of Australia* **15**: 167–175.

Priestley, C. H. B. (1959) Heat conduction and temperature profiles in air and soil. *Journal of the Australian Institute of Agricultural Science* **25**: 94–107.

Prodon, R., Fons, R. & Athias-Binche, F. (1987) The impact of fire on animal communities in Mediterranean area. In *The Role of Fire in Ecological Systems*, ed. Trabaud, L., pp. 121–157. The Hague: SPB Academic.

Purdie, R. W. (1977) Early stages of regeneration after burning in dry sclerophyll vegetation. II. Regeneration by seed germination. *Australian Journal of Botany* **25**: 35–46.

Purdie, R. W. & Slatyer, R. O. (1976) Vegetation succession after fire in sclerophyll woodland communities in south-eastern Australia. *Australian Journal of Ecology* **1**: 223–236.

Pyke, G. H. (1983) Relationship between time since the last fire and flowering in *Telopea speciosissima* R. Br. and *Lambertia formosa* Sm. *Australian Journal of Botany* **31**: 293–296.

Pyne, S. J. (1982) *Fire in America: A Cultural History of Wildland and Rural Fire*. Princeton: Princeton University Press.

Pyne, S. J. (1984) *Introduction to Wildland Fire*. New York: Wiley & Sons.

Pyne, S. J. (1991a) Fire down under: how the first Australians put a continent to the torch. *The Sciences*, **March/April**: 39–45.

Pyne, S. J. (1991b) *Burning Bush: A Fire History of Australia*. New York: Holt.

Quinlivan, B. J. (1971) Seed coat impermeability in legumes. *Journal of the Australian Institute of Agricultural Science* **37**: 283–295.

Quinn, R. D. (1986) Mammalian herbivory and resilience in Mediterranean-climate ecosystems. In *Resilience in Mediterranean-type Ecosystems*, ed. Dell, B., Hopkins, A. J. M. & Lamont, B. B., pp. 113–128. Dordrecht: Junk.

Raison, R. J., Woods, P. V. & Khanna, P. K. (1986) Decomposition and accumulation of litter after fire in sub-alpine eucalypt forests. *Australian Journal of Ecology* **11**: 9–19.

Raup, H. M. (1981) Physical disturbance in the life of plants. In *Biotic Crises In Ecological and Evolutionary Time*, ed. Nitecki, M. H., pp. 39–52. New York: Academic Press.

Recher, H. F. & Christensen, P. E. S. (1981) Fire and the evolution of the Australian biota. In *Ecological Biogeography of Australia*, ed. Keast, A., pp. 137–162. The Hague: Junk.

Recher, H. F., Allen, D. & Gowing, G. (1985) The impact of wildfire on birds in an intensively logged forest. In *Birds of Eucalypt Forests and Woodlands*, ed. Keast, A., Recher, H. F., Ford, H. & Saunders, D., pp. 283–290. Sydney: Surrey Beatty.

Recher, H. F., Shields, J., Kavanah, R. & Webb, G. (1987) Retaining remnant mature forest for nature conservation at Eden, New South Wales: a review of theory and practice. In *Nature Conservation: The Role of Remnants of Native Vegetation*, ed. Saunders, D. A. *et al.*, pp. 177–194. Sydney: Surrey Beatty.

Riba, M. & Terradas, J. (1987) Characteristiques de la resposta als incendis en els ecosistemes mediterranis. In *Ecosistemes Terrestres*, ed. Terredas, J., pp. 63–75. Spain: Diputacio de Barcelona.

Rice, L. A. (1932) The effect of fire on the prairie animal communities. *Ecology* **13**: 392–401.

Rice, S. K. (1993) Vegetation establishment in post-fire *Adenostoma* chaparral in relation to fine-scale pattern in fire intensity and soil nutrients. *Journal of Vegetation Science* **4**: 115–124.

Richardson, R. J. & Holliday, N. J. (1982) Occurrence of carabid beetles (Coleoptera: Carabidae) in a boreal forest damaged by fire. *Canadian Entomologist* **114**: 509–514.

Rickard, W. H. (1970) Ground dwelling beetles in burned and unburned vegetation. *Journal of Rangeland Management* **23**: 293–297.

Riechert, S. E. & Reeder, W. G. (1972) Effects of fire on spider distribution in southwestern Wisconsin prairies. In *Proceedings of the Second Midwest Prairie Conference*, ed. Zimmerman, J. H., pp. 73–90. Madison: University of Wisconsin Arboretum.

Ritchie, J. C. (1959) The vegetation of northern Manitoba. III. Studies in the subarctic. *Arctic Institute of North America Technical Paper No. 3*.

Robbins, L. E. & Myers, R. L. (1992) Seasonal effects of prescribed burning in Florida: a review. *Tall Timbers Research Station, Miscellaneous Publication No. 8*.

Robinson, N. H. (1977) The need for joining Illawarra wilderness areas. *Australian Zoologist* **19**: 125–132.

Rolston, M. P. (1978) Water impermeable seed dormancy. *Botanical Review* **44**: 365–396.

Romme, W. (1980) Fire history terminology: report of the Ad Hoc committee. In *Proceedings of the Fire History Workshop*, ed. Stokes, M. A. & Dieterich, J. H., pp. 135–137. *USDA Forest Service General Technical Report RM-81*.

Rose-Innes, R. (1972) Fire in west African savanna. *Proceedings of the Tall Timbers Fire Ecology Conference* **11**: 147–173.

Rothermel, R. C. (1972) A mathematical model for predicting fire spread in wildland fuels. *USDA Forest Service Research Papers INT-115*.

Rothermel, R. C. & Deeming, J. E. (1980) Measuring and interpreting fire behaviour for correlation with fire effects. *USDA Forest Service General Technical Report INT-93*.

Rowe, J. S. (1983) Concepts of fire effects on plant individuals and species. In *The Role of Fire in Northern Circumpolar Ecosystems*, ed. Wein, R. W. & MacLean, D. A., pp. 135–154. London: Wiley.

Rowe, J. S. & Scotter, G. W. (1973) Fire in the boreal forest. *Quaternary Research* **3**: 444–464.

Rowley, I. & Brooker, M. (1987) The response of a small insectivorous bird to fire in heathlands. In *Nature Conservation: The Role of Remnants of Native Vegetation*, ed. Saunders, D. A., Arnold, G. W., Burbidge, A. A. & Hopkins, A. J. M., pp. 211–218. Sydney: Surrey Beatty.

Rullo, J. C. (1982) Post-fire response in northern sandplain heath species important to bee pastures and fire control management Honours Thesis, University of Western Australia.

Rundel, P. W. (1973) The relationship between basal fire scars and crown damage in Giant Sequoia. *Ecology* **54**: 210–213.

Rundel, P. W. (1982) Fire as an ecological factor. In *Plant Physiological Ecology I: Responses to the Physical Environment*, ed. Lange, O. L., Nobel, P. S., Osmond, C. B. & Ziegler, H., pp. 501–538. Berlin: Springer-Verlag.

Russell, R. P. & Parsons, R. F. (1978) Effects of time since fire on heath floristics at Wilson's Promontory, Southern Australia. *Australian Journal of Botany* **26**: 53–61.

Sabiiti, E. N. & Wein, R. W. (1987) Fire and *Acacia* seeds: a hypothesis of colonization success. *Journal of Ecology* **74**: 937–946.

Scheidlinger, C. R. & Zedler, P. H. (1986) *Response of Torrey Pine to Spring and Winter Controlled Burns*. Final Report on State of California Standard Agreement # 4–400–5076 State Department of Parks and Recreation, San Diego, California.

Schiff, A. L. (1962) *Fire and Water*. Massachussetts: Harvard University Press.

Schlesinger, W. H. & Gill, D. S. (1978) Demographic studies of the chaparral shrub, Ceanothus megacarpus, in the Santa Ynez Mountains, California. *Ecology* **59**: 1256–1263.

Schlesinger, W. H. & Gill, D. S. (1980) Biomass production and changes in the availability of light, water and nutrients during the development of pure stands of the chaparral *Ceanothus megacarpus* after fire. *Ecology* **61**: 781–789.

Schmidt-Nielsen, K. (1979) *Animal Physiology: Adaptation and Environment* (2nd edn.). Cambridge: Cambridge University Press.

Schullery, P. (1989) The fires and fire policy. *Bioscience* **39**: 686–694.

Scott, D. F. & van Wyk, D. B. (1992) The effects of fire on soil water repellency, catchment sediment yields and streamflow. In *Fire in South African Mountain Fynbos: Ecosystem, Community and Species Response at Swartboskloof*, ed. van Wilgen, B. W., Richardson, D. M., Kruger, F. J. & van Hensbergen, H. J., pp. 216–233. Berlin: Springer-Verlag.

Scott, J. K. (1982) The impact of destuctive insects on reproduction in six species of *Banksia* L.f. (Proteaceae). *Australian Journal of Zoology* **30**: 901–921.

Scott, J. K. & Black, R. (1981) Selective predation by White-tailed Black Cockatoos on fruit of *Banksia attenuata* containing the seed-eating weevil *Alphitopis nivea*. *Australian Wildlife Research* **8**: 421–430.

Selkirk, P. M. & Adamson, D. (1981) The effect of fire on Sydney sandstone. In *Bushfires: Their Effect on Australian Life and Landscape*, ed. Stanbury, P., pp. 25–31. Sydney: Macleay Museum, University of Sydney.

Shafi, M. I. & Yarranton, G. A. (1973) Vegetational heterogeneity during a secondary (postfire) succession. *Canadian Journal of Botany* **51**: 73–90.

Shea, S. R., McCormick, J. & Portlock, C. C. (1979) The effect of fires on regeneration of leguminous species in the northern jarrah (*Eucalyptus marginata* Sm) forest of Western Australia. *Australian Journal of Ecology* **4**: 195–205.

Shea, S. R., Peet, G. B. & Cheney, N. P. (1981) The role of fire in forest management. In *Fire and the Australian Biota*, ed. Gill, A. M., Groves, R. H. & Noble, I. R., pp. 443–470. Canberra: Australian Academy of Sciences.

Siddiqi, M. Y., Myerscough, P. J. & Carolin, R. C. (1976) Studies on the ecology of coastal heath in New South Wales. IV. Seed survival, germination, seedling establishment and early growth of *Banksia serratifolia* Salisb., *B. aspleniifolia* Salisb. and *B. ericifolia* L. F. in relation to fire. *Australian Journal of Ecology* 1: 175–183.

Silvertown, J. W. (1982) *Introduction to Plant Population Ecology*. London: Longman.

Singh, G., Kershaw, A. P. & Clark, R. (1981) Quaternary vegetation and fire history in Australia. In *Fire and the Australian Biota*, ed. Gill, A. M., Groves, R. H. & Noble, I. R., pp. 23–54. Canberra: Australian Academy of Science.

Singh, R. S. (1993) Effect of winter fire on primary productivity and nutrient concentration of a dry tropical savanna. *Vegetatio* **106**: 63–71.

Smith, D. W. & Sparling, J. H. (1966) The temperatures of surface fires in Jack Pine Barren. I. The variation in temperature with time. *Canadian Journal of Botany* **44**: 1285–1292.

Smith, G. T. (1985) Fire effects on populations of the noisy scrub-bird, western bristle-bird and western whip-bird. In *Symposium on Fire Ecology and Management in Western Australian Ecosystems*, ed. Ford, J. R., pp. 95–103. Botany Department, University of Western Australia.

Smith, R. E., van Wilgen, B. W., Forsyth, G. G. & Richardson, D. M. (1992) Coexistence of seeders and sprouters in a fire-prone environment: the role of ecophysiology and soil moisture. In *Fire In South African Mountain Fynbos: Ecosystem, Community and Species Response at Swartboskloof*, ed. van Wilgen, B. W., Richardson, D. M., Kruger, F. J. & van Hensbergen, H. J., pp. 108–122. Berlin: Springer-Verlag.

Snyder, J. R. (1984) The role of fire: Mutch ado about nothing? *Oikos* **43**: 404–405.

Snyder, J. R., Herndon, A. & Robertson, Jr. W. B. (1990) South Florida rockland. In *Ecosystems of Florida*, ed. Myers, R. and Ewel, J. J., pp. 230–277. Orlando: Academic Press.

Sonia, L. & Heslehurst, M. R. (1978) Germination characteristics of some *Banksia* species. *Australian Journal of Ecology* **3**: 179–186.

Southwood, T. R. E. (1966) *Ecological Methods*. London: Methuen.

Spalt, K. W. & Reifsnyder, W. E. (1962) *Bark Characteristics and Fire Resistance: A Literature Survey*. USDA Forest Service Southern Forest Experimental Station Occasional Paper 193.

Specht, R. L. (1979) Heathlands and related shrublands of the world. In *Heathlands and Related Shrublands of the World: 9A Descriptive Studies*, ed. Specht, R. L., pp. 1–18. Amsterdam: Elsevier.

Specht, R. L. (1981) Conservation: Australian heathlands. In *Heathlands and Related Shrublands of the World. 9B. Analytical Studies*, ed. Specht, R. L., pp. 235–240. Amsterdam: Elsevier.

Specht, R. L., Rayson, P. & Jackman, M. E. (1958) Dark Island Heath (Ninety Mile Plain, South Australia) VI. Pyric succession: changes in composition, coverage, dry weight and mineral nutrient status. *Australian Journal of Botany* **6**: 59–88.

Springett, J. A. (1971) The effects of fire on litter decomposition and on the soil fauna in a *Pinus pinaster* plantation. IV. Colloquium Pedobiologiae. *Annals of Zoology* **17**: 257–263.

Springett, J. A. (1976) The effect of prescribed burning on the soil fauna and on litter decomposition in Western Australian forests. *Australian Journal of Ecology* 1: 77–82.

Springett, J. A. (1979) The effects of a single hot summer fire on soil fauna and on litter decomposition in jarrah (*Eucalyptus marginata*) forest in Western Australia. *Australian Journal of Ecology* 4: 279–291.

Starker, T. J. (1934) Fire resistance in the forest. *Journal of Forestry* 32: 462–467.

Stern, K. & Roche, L. (1974) *Genetics of Forest Ecosystems*. Berlin: Springer-Verlag.

Stewart, O. C. (1956) Fire as the first great force employed by man. In *Man's Role in Changing the Face of the Earth,* ed. Thomas, W. L., pp. 115–133. Chicago: University of Chicago Press.

Stewart-Oaten, A., Murdoch, W. W. & Parker, K. R. (1986) Environmental impact assessment: pseudoreplication in time? *Ecology* 67: 929–940.

Stinson, K. J. & Wright, H. A. (1969) Temperature and headfires in the southern mixed prairie of Texas. *Journal of Rangeland Management* 22: 169–174.

Stocker, G. C. (1966) Effects of fires on vegetation in the Northern Territory. *Australian Forestry* 30: 223–230.

Stocker, G. C. and Mott J. J. (1981) Fire in the tropical forests and woodlands of northern Australia. In *Fire and the Australian Biota*, ed. Gill, A. M., Groves, R. H. & Noble, I. R., pp. 427–439. Canberra: Australian Academy of Science.

Stocker, G. C. & Sturtz, J. D. (1966) The use of fire to establish Townsville lucerne in the Northern Territory. *Australian Journal of Experimental Agriculture and Animal Husbandry* 6: 277–279.

Stoddard, H. L. (1935) Use of controlled fire in southeastern upland game management. *Journal of Forestry* 33: 346–351.

Stokes, M. A. & Dieterich, J. H. (eds.) (1980) *Proceedings of the Fire History Workshop.* USDA Forest Service General Technical Report RM-81.

Strelein, G. J. (1988) Gum leaf skeletoniser moth, *Uraba lugens*, in the forests of Western Australia. *Australian Forestry* 51: 197–204.

Swanson, F. J. (1981) Fire and geomorphic processes. In *Proceedings of a Conference on Fire Regimes and Ecosystem Properties*, pp. 401–420. USDA Forest Service General Technical Report WO-26.

Sweeney, J. R. (1956) Responses of vegetation to fire. *University of California Publications in Botany* 28: 143–250.

Sweeney, J. R. (1968) Ecology of some fire types in northern California. *Proceedings of the Tall Timbers Fire Ecology Conference* 7: 111–126.

Symonides, E. (1979) The structure of population dynamics of psammophytes on inland dunes. II. Loose-sod populations. *Ekologia Polska* 27: 191–234.

Takahashi, F. (1982) Research project in burnt forest at Etajima Island. In *Researches Related to the UNESCO's Man and the Biosphere Programme in Japan, 1981–1982,* pp. 25–34. MAB, Tokyo, Japan.

Tamm, J. C. (1986) Fünfjährige Collembolensukzession auf einem verbrannten Kiefernwalboden in Niedersachen, BRD.(Five-year post-fire succession of the surface-dwelling Collembola inhabiting the floor of a pine forest in Niedersachen, FRG.) *Pedbiologia* 29: 113–127.

Tap, P. M. & Whelan, R. J. (1984) The effect of fire on populations of heathland invertebrates. In *Medecos IV*, ed. Dell, B., pp. 147–148. Botany Department,

University of Western Australia.

Taylor, R. J. (1991) Short note. Plants, fungi and bettongs: a fire-dependent co-evolutionary relationship? *Australian Journal of Ecology* **16**: 409–411.

Taylor, R. J. (1992) Reply. Fire, mycorrhizal fungi and management of mycophagous marsupials. *Australian Journal of Ecology* **17**: 227–228.

Teich, A. H. (1970) Cone serotiny and inbreeding in natural populations of *Pinus banksiana* and *Pinus contorta*. *Canadian Journal of Botany* **48**: 1805–1809.

Tester, J. R. & Marshall, W. H. (1961) *A Study of Certain Plant and Animal Interactions on a Native Prairie in Northwestern Minnesota*. Minnesota Museum of Natural History, Occasional Paper No. 8. Minneapolis: University of Minnesota Press.

Tevis, Jr, L. (1956) Effect of slash and burn on forest mice. *Journal of Wildlife Management* **20**: 405–409.

Thompson, K. (1978) The occurrence of buried viable seeds in relation to environmental gradients. *Journal of Biogeography* **5**: 425–430.

Trabaud, L. (1974) Experimental study of the effects of prescribed burning on a *Quercus coccifera* L. garrigue. *Proceedings of the Tall Timbers Fire Ecology Conference* **13**: 97–129.

Trabaud, L. (1979) Etude du comportement du feu dans la Garrique de Chenes kermes a partir des temperatures et des vitesses de propagation. *Annales des Sciences forestières* **36**: 13–38.

Trabaud, L. (1980) *Impact biologique et écologique des feux de végétation sur l'organisation, la structure et l'évolution de la végétation des garrigues du Bas-Languedoc*. Thèse Doctoral Etat, Université des Sciences et Techniques du Languedoc, Montpellier.

Trabaud, L. (1987a) Fire and the survival traits of plants. In *The Role of Fire in Ecological Systems*, ed. Trabaud, L., pp. 65–89. Hague: SPB Academic.

Trabaud, L. (ed.) (1987b) *The Role of Fire in Ecological Systems*. The Hague: SPB Academic.

Trabaud, L. & Lepart, J. (1980) Diversity and stability in Garrigue ecosystems after fire. *Vegetatio* **43**: 49–57.

Trabaud, L. & Lepart, J. (1981) Changes in the floristic composition of a *Quercus coccifera* L. garrigue in relation to different fire regimes. *Vegetatio* **46**: 105–116.

Trabaud, L., Grosman, J. & Walter, T. (1985) Recovery of burnt *Pinus halepensis* Mill. forests. I. Understorey and litter phytomass development after wildfire. *Forest Ecology and Management* **12**: 269–277.

Trollope (1984) Fire in savanna. In *Ecological Effects of Fire in South African Ecosystems*, ed. Booysen, P.de V. & Tainton, N. M., pp. 151–175. Berlin: Springer-Verlag.

Twigg, L. E., Fox, B. J. & Jia, L. (1989) The modified primary succession following sand mining: a validation of the use of chronosequence analysis. *Australian Journal of Ecology* **14**: 441–447.

Uggla, E. (1958) *Ecological Effects of Fire on North Swedish Forests*. Stockholm: Almqvist & Wiksell.

Uhl, C. & Kauffman, J. B. (1990) Deforestation, fire susceptability and potential tree responses to fire in the eastern Amazon. *Ecology* **71**: 437–449.

Underwood, A. J. (1986) The analysis of competition by field experiments. In *Community Ecology: Pattern and Process*, ed. Kikkawa, J. & Anderson, D. J., pp. 240–268. Melbourne: Blackwell Scientific.

van Amburg, G. L., Swaby, J. A. & Pemble J. H. (1981) Response of arthropods to a spring burn of a tallgrass prairie in northwestern Minnesota. In *Proceedings of the 6th North American Prairie Conference – The Prairie Peninsula: In the Shadow of Transeau*, ed. Stuckey, R. L. & Reese, K. J., pp. 240–243. Ohio Biological Survey Notes 15.

van Cleve, K. & Viereck, L. A. (1981) Forest succession in relation to nutrient cycling in the boreal forest of Alaska. In *Forest Succession: Concepts and Applications*, ed. West, D. C., Shugart, H. H. & Botkin, D. B., pp. 185–211. New York: Springer-Verlag.

van Hensbergen, H. J., Botha, S. A., Forsyth, G. G. & Le Maitre, D. C. (1992) Do small mammals govern vegetation recovery after fire in fynbos? In *Fire In South African Mountain Fynbos: Ecosystem, Community and Species Response at Swartboskloof*, ed. van Wilgen, B. W., Richardson, D. M., Kruger, F. J. & van Hensbergen, H. J., pp. 182–202. Berlin: Springer-Verlag.

van Wagner, C. E. (1990) Six decades of forest fire science in Canada. *Forestry Chronicle*, **April**: 133–137.

van Wilgen, B. W. (1986) A simple relationship for estimating the intensity of fires in natural vegetation. *South African Journal of Botany* 52: 384–385.

van Wilgen, B. W., Everson, C. S. & Trollope, W. S. W. (1990) Fire management in southern Africa: some examples of current objectives, practices and problems. In *Fire in the Tropical Biota: Ecosystem Processes and Global Challenges*, ed. Goldammer, J. G., pp. 179–215. Berlin: Springer-Verlag.

van Wilgen, B. W., Bond, W. J. & Richardson, D. M. (1992a) Ecosystem management. In *The Ecology of Fynbos: Nutrients, Fire and Diversity*, ed. Cowling, R., pp. 345–371. Oxford: Oxford University Press.

van Wilgen, B. W., Richardson, D. M., Kruger, F. J. & van Hensbergen, H. J. (eds.) (1992b) *Fire in South African Mountain Fynbos*. Berlin: Springer-Verlag.

Varley, G. C., Gradwell, G. R. & Hassell, M. P. (1973) *Insect Population Ecology*. Oxford: Blackwell Scientific.

Veno, P. A. (1976) Successional relationships of five Florida plant communities. *Ecology* 57: 498–508.

Viereck, L. A. (1973) Wildfire in the taiga of Alaska. *Quaternary Research* 3: 465–495.

Viereck, L. A. & Dyrness, C. T. (eds.) (1979) Ecological effects of the Wickersham Dome fire near Fairbanks, Alaska. *USDA Forest Service General Technical Report PNW-90*.

Vines, R. G. (1981) Physics and chemistry of rural fires. In *Fire and the Australian Biota*, ed. Gill, A. M., Groves, R. H. & Noble, I. R., pp. 129–149. Canberra: Australian Academy of Science.

Viro, P. J. (1974) Effects of forest fire on soil. In *Fire and Ecosystems*, ed. Kozlowski, T. T. & Ahlgren, C. E., pp. 7–46. New York: Academic Press.

Vlahos, S. & Bell, D. T. (1986) Soil seed bank components of the northern jarrah forest of Western Australia. *Australian Journal of Ecology* 11: 171–179.

Vogl, R. J. (1969) The role of fire in the evolution of the Hawaiian flora and vegetation. *Proceedings of the Tall Timber Fire Ecology Conference* 9: 5–60.

Vogl, R. J. (1973) Effects of fire on the plants and animals of a Florida wetland. *American Midland Naturalist* 89: 334–347.

Vogl, R. J. & Schorr, P. K. (1972) Fire and manzanita chaparral in the San Jacinto mountains, California. *Ecology* 53: 1179–1188.

Wade, D. D. (1991) High-intensity prescribed fire to maintain spartia marsh at the urban–wildland interface. *Proceedings of the Tall Timbers Fire Ecology Conference* **17**: 211–216.

Wade, D. D., Ewel, J. J. & Hofsetter, R. (1980) Fire in south Florida ecosystems. *USDA Forest Service General Technical Report SE-17.*

Wahlenberg, W. G. (1946) *Longleaf Pine: It's Use, Ecology, Regeneration, Protection, Growth and Management.* Washington DC: Charles Lathrop Pack Forest Foundation.

Waldrop, T. A., Van Lear, D. H., Lloyd, F. T. & Harms, W. R. (1987) Long-term studies of prescribed burning in loblolly pine forests of the Southeastern Coastal Plain. *USDA Forest Service General Technical Report SE-45.*

Walker, D. (1982) The development of resilience in burned vegetation. In *The Plant Community as a Working Mechanism*, ed. Newman, E. I., pp. 27–43. London: Blackwell Scientific.

Walker, J. S. (1963) *Bushfire Combustion Studies – Fuel Pyrolysis.* CSIRO Australian Chemistry Research Laboratories, Mimeographed Report.

Walker, J. S. (1981) Fuel dynamics in Australian vegetation. In *Fire and the Australian Biota*, ed. Gill, A. M., Groves, R. H. & Noble, I. R., pp. 101–128. Canberra: Australian Academy of Science.

Wallace, M. M. H. (1961) Pasture burning and its effect on the aestivating eggs of *Halotydeus destructor* (Tuck.). *Australian Journal of Experimental Agriculture and Animal Husbandry* **1**: 109–111.

Wallace, W. R. (1966) Fire in the jarrah forest environment. *Journal of the Royal Society of Western Australia* **49**: 33–44.

Walstad, J. D., Radosevich, S. R. & Sandberg, D. V. (eds.) (1990) *Natural and Prescribed Fire in Pacific Northwest Forests.* Corvallis: Oregon State University Press.

Wardrop, A. B. (1983) The opening mechanism of follicles of some species of *Banksia. Australian Journal of Botany* **31**: 485–500.

Warren, R. & Fordham, A. J. (1978) The fire pines. *Arnoldia* **38**: 1–11.

Warren, S. D., Scifres, C. J. & Teel, P. D. (1987) Response of grassland arthropods to burning: a review. *Agriculture, Ecosystems and Environment* **19**: 105–130.

Watt, A. S. (1947) Pattern and process in the plant community. *Journal of Ecology* **35**: 1–22.

Wein, R. W. & MacLean, D. A. (eds.) (1983) *The Role of Fire in Northern Circumpolar Ecosystems.* New York: Wiley.

Wellington, A. B. & Noble, I. R. (1985a) Post-fire recruitment and mortality in a population of the mallee *Eucalyptus incrassata* in semi-arid south-eastern Australia. *Journal of Ecology* **73**: 645–656.

Wellington, A. B. & Noble, I. R. (1985b) Seed dynamics and factors limiting recruitment of the mallee *Eucalyptus incrassata* in semi-arid south-eastern Australia. *Journal of Ecology* **73**: 657–666.

Wells, C. G., Campbell, R. E., DeBano, L. F., Lewis, C. E., Frederickson, R. L., Franklin, E. C., Froelich, R. C. & Dunn, P. H. (eds.) (1979) *Effects of Fire on Soil. USDA Forest Service General Technical Report WO-7.*

Wells, P. V. (1969) The relation between mode of reproduction and extent of speciation in woody genera of the Californian chaparral. *Evolution* **23**: 264–267.

West, O. (1965) *Fire in Vegetation and its Use in Pasture Management, with Special*

Reference to Tropical and Subtropical Africa. Commonwealth Agriculture Bureau, Mimeographed Publication 1/1965, Hurley, Berkshire.

Westman, W. E. (1986) Resilience: concepts and measures. In *Resilience in Mediterranean-type Ecosystems*, ed. Dell, B., Hopkins, A. J. M. & Lamont, B. B., pp. 5–19. Dordrecht: Junk.

Westoby, M. (1981) How diversified seed germination behaviour is selected. *American Naturalist* **118**: 882–885.

Westoby, M. (1991) On long-term ecological research in Australia. In *Long-term Ecological Research: An International Perspective*, ed. Risser, P. G., pp. 191–209. New York: Wiley.

Whelan, R. J. (1977) The influence of insect grazers on the establishment of post-fire plant populations. Unpublished PhD Thesis, University of Western Australia.

Whelan, R. J. (1985) Patterns of recruitment of plant populations after fire in Western Australia and Florida. *Proceedings of the Ecological Society of Australia* **14**: 169–178.

Whelan, R. J. (1986) Seed dispersal in relation to fire. In *Seed Dispersal*, ed. D. R. Murray, pp. 237–271. Sydney: Academic Press.

Whelan, R. J. (1989) The influence of fauna on plant species composition. In *Animals in Primary Succession: The Role of Fauna in Reclaimed Lands*, ed. Majer, J. D., pp. 107–142. Cambridge: Cambridge University Press.

Whelan, R. J. & Main, A. R. (1979) Insect grazing and post-fire plant succession in south-west Australian woodland. *Australian Journal of Ecology* **4**: 387–398.

Whelan, R. J. & Muston, R. M. (1991) Fire regimes and management in southeastern Australia. *Proceedings of the Tall Timbers Fire Ecology Conference* **17**: 235–258.

Whelan, R. J. & Tait, I. (1995) Responses of plant populations to fire: fire season as an understudied element of fire regime. *CalmScience* supplement **4**: 147–150.

Whelan, R. J., Langedyk, W. & Pashby, A. S. (1980) The effects of wildfire on arthropod populations in jarrah–*Banksia* woodland. *Western Australian Naturalist* **14**: 214–220.

White, J. (1980) Demographic factors in populations of plants. In *Demography and Evolution in Plant Populations*, ed. Solbrig, O. T., pp. 21–48. Oxford: Blackwell Scientific.

Whittaker, R. H. (1975) *Communities and Ecosystems* (2nd edn.). New York: Macmillan.

Wicht, C. L. (1948) A statistically designed experiment to test the effects of burning on a sclerophyll shrub community. I. Preliminary account. *Transactions of the Royal Society of South Africa* **31**: 479–501.

Wicklow-Howard, M. (1989) The occurrence of vesicular–arbuscular mycorrhizae in burned areas of the Snake River Birds of Prey Area, Idaho. *Mycotaxon* **34**: 253–257.

Williams, D. T. (1972) *Smoke at Palm Beach during the 1971 Everglades wildfires*. USDA Forestry Service, Southeast Forest Experiment Station, Macon, GA.

Williams, G. C. (1966) *Adaptation and Natural Selection: A Critique of some Current Evolutionary Thought*. Princeton: Princeton University Press.

Williams, J. E., Whelan, R. J. & Gill, A. M. (1994) Fire and environmental heterogeneity in southern temperate ecosystems: implications for management. *Australian Journal of Botany* **42**: 125–137.

Williamson, G. B. & Black, E. M. (1981) High temperatures of forest fires under pines: a selective advantage over oaks. *Nature* **293**: 643–644.

Wilson, R. E. & Rice, E. L. (1968) Allelopathy as expressed by *Helianthus annuus* and its role in old-field succession. *Bulletin of the Torrey Botanical Club* **95**: 432–448.

Wink, R. L. & Wright, H. A. (1973) Effects of fire on an Ashe juniper community. *Journal of Rangeland Management* **26**: 326–329.

Wirtz, W. O. (1977) Vertebrate post-fire succession. In *Proceedings of the Symposium on the Environmental and Fuel Management in Mediterranean Ecosystems*, pp. 46–57. USDA Forest Service General Technical Report WO-3.

Wirtz, W. O. (1979) Effects of fire on birds in chaparral. In *CAL-NEVA Wildlife, Transactions of the 10th Annual Meeting, Western Section*, pp. 114–124. The Wildlife Society & American Fisheries Society.

Wirtz, W. O. (1982) Post-fire community structure of birds and rodents in southern Californian chaparral. In *Dynamics and Management of Mediterranean-type Ecosystems*, ed. Conrad, C. E. & Oechel, W. C., pp. 241–246. *USDA Forest Service General Technical Report PSW-58*.

Woinarski, J. C. Z. (1990) Effects of fire on the bird communities of tropical woodlands and open forests in northern Australia. *Australian Journal of Ecology* **15**: 1–22.

Wright, H. A. (1980) The role and use of fire in the semidesert grass–shrub type. *USDA Forest Service General Technical Report INT 85*.

Wright, H. A. & Bailey, A. W. (1982) *Fire Ecology: United States and Southern Canada*. New York: Wiley-Interscience.

Wrigley, J. W. & Fagg, M. (1979) *Australian Native Plants*. Melbourne: Collins.

Young, J. A. & Evans, R. A. (1978) Population dynamics after wildfires in sagebrush grasslands. *Journal of Rangeland Management* **31**: 283–289.

Young, J. A. & Evans, R. A. (1985) Demography of Bromus tectorum in *Artemisia* communities. In *The Population Structure of Vegetation*, ed. White, J., pp. 489–502. Dordrecht: Junk.

Zammit, C. (1988) Dynamics of resprouting in the lignotuberous shrub *Banksia oblongifolia*. *Australian Journal of Ecology* **13**: 311–320.

Zammit, C. A. & Hood, C. H. (1986) Impact of flower and seed predators on seed set in two *Banksia* shrubs. *Australian Journal of Ecology* **11**: 187–193.

Zammit, C. & Westoby, M. (1987) Population structure and reproductive status of two *Banksia* shrubs at various times after fire. *Vegetatio* **70**: 11–20.

Zammit, C. & Westoby, M. (1988) Pre-dispersal seed losses, and the survival of seeds and seedlings of two serotinous *Banksia* shrubs in burnt and unburnt heath. *Journal of Ecology* **76**: 200–214.

Zedler, P. H. (1977) Life-history attributes of plants and the fire cycle: a case study in chaparral dominated by *Cupressus forbesii*. In *Proceedings of a Symposium on the Environmental Consequences of Fire and Fuel Management in Mediterranean Ecosystems*, ed. Mooney, H., pp. 451–458. USDA Forest Service General Technical Report WO-3.

Zedler, P. H., Gautier, C. R. & McMaster, G. S. (1983) Vegetation change in response to extreme events: the effect of a short interval between fires in California chaparral and coastal scrub. *Ecology* **64**: 809–818.

Zikria, B. A., Weston, G. C., Chodoff, M. & Ferrer, J. M. (1972) Smoke and carbon monoxide poisoning in fire victims. *Journal of Trauma* **12**: 641–645.

Index

Wildfires kill many animals, but are *populations* of animals affected? How do animals survive the passage of fire? Why do some tree species survive and others die in a fire? Do frequent fires cause changes in plant community composition? How important is long-distance seed dispersal in vegetation recovery after fire? How does fire affect plant–herbivore interactions and predator–prey interactions? What are the effects of frequently applied, out of season fires for land management?

Answering questions such as these requires an understanding of the ecological effects of fire. Aimed at senior undergraduate students, researchers, foresters and other land managers, Professor Whelan's book examines the changes wrought by fires with reference to general ecological theory. The impact of fires on individual organisms, populations and communities are examined separately, and emphasis is placed on the importance of fire regime. Each chapter includes a listing of 'outstanding questions' that identify gaps in current knowledge. The book finishes by summarising the major aspects of ecology that are of particular relevance to management of fires – both protection against wildfires and deliberate use of fire.

Cover illustration by Donald Keys

Cover design by Chris McLeod and Andy Wilson

ISBN 0-521-33814-X

CAMBRIDGE
UNIVERSITY PRESS

9 780521 338141